The Public Management of Forestry Projects

ORGANISATION FOR ECONOMIC CO-OPERATION AND DEVELOPMENT

Pursuant to article 1 of the Convention signed in Paris on 14th December, 1960, and which came into force on 30th September, 1961, the Organisation for Economic Co-operation and Development (OECD) shall promote policies designed:

- to achieve the highest sustainable economic growth and employment and a rising standard of living in Member countries, while maintaining financial stability, and thus to contribute to the development of the world economy;
- to contribute to sound economic expansion in Member as well as non-member countries in the process of economic development; and
- to contribute to the expansion of world trade on a multilateral, non-discriminatory basis in accordance with international obligations.

The Signatories of the Convention on the OECD are Austria, Belgium, Canada, Denmark, France, the Federal Republic of Germany, Greece, Iceland, Ireland, Italy, Luxembourg, the Netherlands, Norway, Portugal, Spain, Sweden, Switzerland, Turkey, the United Kingdom and the United States. The following countries acceded subsequently to this Convention (the dates are those on which the instruments of accession were deposited): Japan (28th April, 1964), Finland (28th January, 1969), Australia (7th June, 1971) and New Zealand (29th May, 1973).

The Socialist Federal Republic of Yugoslavia takes part in certain work of the OECD (agreement of 28th October, 1961).

Publié en français sous le titre:

LA GESTION PUBLIQUE
DES PROJETS FORESTIERS

Photo Cover: Information CANADA No. 781926 NANAIMO Lakes BC

This book results from a project carried out jointly by sixteen OECD Member countries. Each of these nominated a national team of experts to participate in the work.

The teams represented the range of governmental institutions involved in the planning and management of forestry resources. They met regularly to exchange views and review the present document in the light of practice in their respective countries.

They were guided by an interdisciplinary team of experts composed of: Mr. Fernando Barrientos and the late Mr. Lorenzo Ferrer-Martin, National Institute for Conservation of Nature (Spain); Professor Hans Gregersen, University of Minnesota (USA); Mr. Jean de Montgolfier, Ministry of Agriculture (France); Mr. Jorgen Randers, Dean of the Norvegian School of Management, in his former capacity as Director, Resource Policy Group (Norway).

This book is published on the responsibility of the Secretary General of OECD.

Also available

THE STATE OF THE ENVIRONMENT – 1985 (June 1985)
(97 85 04 1) ISBN 92-64-12713-5 272 pages £16.50 US$33.00 F165.00 DM73.00

MANAGEMENT OF WATER PROJECTS. Decision-making and Investment Appraisal (May 1985)
(42 85 01 1) ISBN 92-64-12695-3 254 pages £12.00 US$24.00 F120.00 DM53.00

STRATEGIES FOR CHANGE AND REFORM IN PUBLIC MANAGEMENT (September 1980)
(42 80 01 1) ISBN 92-64-12121-8 242 pages £7.10 US$16.00 F64.00 DM32.00

To be published

RURAL PUBLIC MANAGEMENT

Prices charged at the OECD Bookshop.

THE OECD CATALOGUE OF PUBLICATIONS and supplements will be sent free of charge on request addressed either to OECD Publications Service, Sales and Distribution Division, 2, rue André-Pascal, 75775 PARIS CEDEX 16, or to the OECD Sales Agent in your country.

TABLE OF CONTENTS

PREFACE 7

Chapter I. INTRODUCTION 8

A. What is Project Appraisal 8
B. How Project Appraisals are Organised 11
C. The Goal of this Handbook 12

Chapter II. THE APPRAISAL PROCESS 15

A. The Main Stages in Project Appraisal 15
B. Some Basic Principles of Project Appraisal 18
C. Presentation of the Project Appraisal 24

Chapter III. SOCIAL EFFECTS 28

A. Questions Decision-makers Need Answers to 29
B. Procedure for Answering the Questions 29
C. Procedure During Each Stage of the Appraisal Process 30
D. A Sample Case: The Inner Valley Road Project 34

Chapter IV. ENVIRONMENTAL EFFECTS 39

A. Questions Decision-makers Need Answers to 39
B. Procedure During Each Stage of the Appraisal Process 40
C. Examples of Environmental Effects 42

Chapter V. ECONOMIC AND FINANCIAL EFFECTS 47

A. Questions Decision-makers Need Answers to 47
B. Procedure for Answering the Questions 48
C. Procedure at Each Stage of the Appraisal Process 57
D. Summary 64

Chapter VI. A CASE STUDY: THE MYTHOPIA WATERSHED PROJECT 66

A. How the Project Idea was Conceived 66
B. Organising the Stage I Assessment 67
C. Moving Ahead on Stage II Assessment 77
D. The Stage III Assessment: Preparing the Final Report 80

APPENDICES

I. TECHNICAL CONSIDERATIONS: FINANCIAL AND ECONOMIC EFFECTS 83
II. TAKING ACCOUNT OF RENEWABLE NATURAL RESOURCES IN ECONOMIC CALCULATIONS 89

GLOSSARY 96

BIBLIOGRAPHY 99

TABLES

1.1.	Common Effects of Forestry Investment Projects	9
1.2.	Examples of "Routine" and "One-time" Projects and Decisions	10
2.1.	Description of Appraisal Stages	17
2.2.	A Hypothetical Case: Effects of Project Alternatives	22
2.3.	A Hypothetical Case: Additional Project Alternatives	24
3.1.	Checklist for Common Social Effects of Forestry Investment Projects	31
3.2.	Discussion Group Ranking of Various Project Alternatives	36
3.3.	Social Effects of Various Project Alternatives	36
3.4.	Community Ranking of Project Alternatives	38
4.1.	Relative Effects of Watershed Activities on Water Yield Characteristics	44
5.1.	Project Physical Inputs and Outputs	50
5.2.	Forest Agency – Inputs and Outputs	50
5.3.	Private Landowners – Inputs and Outputs	50
5.4.	Additional Direct Employment by Income Groups	51
5.5.	Imported Inputs and Exported Outputs at Different Times	51
5.6.	Estimated Market Prices	51
5.7.	Shadow Prices	52
5.8.	Total Economic Value Flow Table	52
5.9.	Forest Agency Outlays	53
5.10.	Cash Flow Table for Private Landowners	53
5.11.	Additional Income by Income Groups	53
5.12.	Balance of Payments Effects	55
5.13.	Examples of Forestry Projects Involving Both Public and Private Participation	56
5.14.	Procedure at Each Stage of the Appraisal Process	58
5.15.	Expected Financial Rate of Return (FRR) by Farmer Income/Site Class Groups	62
6.1.	Watershed Project: Identification of Water Benefits	68
6.2.	Areas Associated with Each Project Component	69
6.3.	Inputs and Outputs – Production Forest Component	70
6.4.	Value Flow Table	72
6.5.	Sensitivity Analyses	76

FIGURES

2.1.	Stages in the Project Planning Process	16
2.2.	How to Show a Quantified Variable Overtime	25
2.3.	How to Present a Split Quantified Variable	25
2.4.	A Hypothetical Case: Comparison of Project Alternatives	26
2.5.	A Hypothetical Case: Ranking of Four Alternatives by Social Groups	26
3.1.	Time Path of Forestry Employment for Various Project Alternatives	37
5.1.	Measures of Project Worth: Net Present Worth and Internal Rate of Return	49
5.2.	The Procedure in Economic and Financial Assessment	63

PREFACE

The contents of this book are the outcome of an activity undertaken in response to the substantial interest shown by many OECD Member countries, and other international organisations in the management of forestry as a vital renewable natural resource.

Whilst there is an increasing demand for forestry products, there is also growing recognition that forestry is a limited economic resource, that it plays a major role in maintaining rural communities, and that it is an important factor in the preservation of soil and water resources and natural beauty.

Management of forest resources involves several critical aspects of public management. Choice of investments in forestry resources provides a significant example of a field where public decision-taking processes and issues are complex. Forestry is a long-term process and it is for the public authorities to avoid shortages by ensuring that investment planning is carried out with sufficient forward lead time. Simultaneously, there is also an increasing demand for careful management and scrutiny of public expenditure.

The objective of the present study is to provide a tool for improvement of forestry project appraisal investment procedures decision-making by including economic, financial, social and environmental factors. It proposes a set of guidelines for assessing forestry projects, and describes in broad terms the sequence of steps for the appraisal process. It also indicates the broad spectrum of effects that may be found relevant in any one project, and tentatively suggests how these various effects can be identified, described and evaluated. The level of detail involved, and the emphasis on any particular step compared to others will differ from situation to situation. Furthermore, account naturally has to be taken of the institutional structures in a country and the national economic and social objectives within which forestry projects are designed, evaluated and selected.

The study covers a broad range of situations and problems of relevance to Member countries and to many non-Member countries. For this reason, the experts appointed to the Steering Committee, set up to oversee the activity expressed the wish that it should be published to ensure its wide circulation. They recommended that it be drawn to the attention of economic and planning agencies, public forestry bodies, and relevant training and research institutions.

As Chairman of the meetings of the OECD Joint Activity on the Management of Forestry Resources, I would like to offer my thanks to all those who co-operated constructively in producing this book, and particularly the consultants, the members of the national teams and the OECD Technical Co-operation Service and the Agricultural and Environment Directorates.

Hans Kr. Seip
General Director of Forestry,
Norway

Chapter I

INTRODUCTION

Before examining the major stages in project appraisal for forestry management, it is useful to explore some basic questions first. What is a project appraisal, and what does it consist of? How is a project appraisal organised? Once these questions have been discussed, it will be easier to show why this study was undertaken and how we feel it can be useful to the OECD countries.

A. What is Project Appraisal?

For our purpose here, a "project" can be roughly defined as any set of interrelated actions that are relevant to and typically take the time of decision-makers. A project appraisal can apply equally well to a large-scale national programme or a small, area-specific project, the difference being mainly in the amount of empirical content included in the appraisal.

As we use the term, a project appraisal is a study undertaken to clarify the probable effects of a proposed project, that is, it attempts to investigate ahead of time the economic, financial, social, environmental and other consequences of a proposed action. Ideally, the final output is a broad yet incisive scenario that depicts likely developments.

As the appraisal is designed to help the decision-maker choose whether or not to implement a project, it must also show what the future would be like if the project were not implemented. In other words, the appraisal team should come up with at least two scenarios: one that describes developments "with the project" and another that describes the future "without the project". The rationale for either going ahead with the project or rejecting it will be based precisely on the differences between the two scenarios.

In practice, though, a project can usually be implemented in more ways than one; there are, in other words, several alternatives to reach the objectives outlined in the project. To facilitate the choice between these alternatives, one would ideally like to have a scenario developed for each one, but this is not always possible, given the constraints of time and money. In any case, it is not the job of the appraisal team to recommend one or the other alternative; rather, it is up to the decision-maker to choose, and to do that, he or she needs the clearest picture possible of the various alternatives and their effects.

It would, of course, be highly desirable to have all future effects of each alternative summed up in a figure or table (see Table 1.1 for a list of some of the most common effects). This is hardly, if ever possible. The decision-maker must usually be content with multi-dimensional scenarios of the diverse effects, even if they do force him to compare "apples and pears" when choosing between project alternatives. Moreoever, no amount of technical work, statistical manipulation or mathematical weighting can spare him from

Table 1.1. COMMON EFFECTS OF FORESTRY INVESTMENT PROJECTS*

Economic and Financial Effects

- Regional and national level of production
- Allocation of resources
- Regional and national income
- National balance of payments
- Stability of income over time
- Distribution of income (both interpersonal and intertemporal)
- Public budgets

Environmental Effects

- Ecological diversity
- Watershed stability
- Wildlife protection
- Soil protection
- Landscape aesthetics
- Natural resource conservation
- National patrimony

Social Effects

- Regional employment
- Working conditions
- Public participation
- Migration flows
- Cultural traditions
- National vulnerability
- Political stability

* For the sake of convenience, we have divided the effects into three categories: environmental, social and economic/financial effects. In point of fact, there is a great deal of overlap between the categories.

making the basically political choice between, say, cheaper wood and fewer jobs. What the appraisal does do is to help the decision-maker see the options more clearly and, whenever possible, the magnitude of the trade-offs involved in the various alternatives.

As we define the appraisal process, it does not include the set of actions leading up to the decision "that a decision needs to be made". Ideally, the project has been defined and the technically feasible project alternatives have been identified before the appraisal process starts. Nor does it include the actual process of making the final selection and deciding whether or not to implement the selected alternative.

As we define it then, the appraisal process involves:

- identification of the effects of the project alternatives;
- description of the effects, either in words or, whenever possible, in quantitative measurements;
- evaluation of the alternatives, by comparing the measures with selected criteria or by determining how the alternatives are ranked by the various social groups involved[1].

In actual practice, however, neither the decision-making process nor the appraisal process is quite so neat and systematic. Though investigations to identify alternatives should normally be a separate activity preceeding appraisals, the appraisal process in fact involves both appraisal and redesign of alternatives. After a first-stage appraisal of the initial alternatives, new ones are generated, then appraised, and so on. Similarly, the decision-making associated with solving a problem or exploiting an opportunity is also an iterative process that takes place in successive stages.

Most institutions involved in forestry have to make two distinct kinds of decision: routine decisions concerning the management and utilisation of public and private forests and decisions for "one time" projects (see Table 1.2 for examples of each). Although an *a priori* appraisal can in principle be applied equally well to assist in both types of decision-making, full-scale appraisals are not usually used for routine decisions.

Table 1.2. EXAMPLES OF "ROUTINE" AND "ONE-TIME" PROJECTS AND DECISIONS

Routine Decisions

- Level of maintenance of roads and other facilities;
- Choice of areas and activities for expenditure of annual timber stand improvement budget;
- Allocation of an annual forest protection budget between fire, insect and disease control;
- Choice of recipients for an existing subsidy scheme.

One-time Decisions

- Construction of a new road to open up a forest area;
- Creation of a new forest park with associated facilities;
- Establishment of a wood-processing facility;
- Implementation of a regional watershed protection project;
- Implementation of a new financial support system;
- Establishment of a nature preserve.

In most countries, routine decisions are based on fairly straightforward "rules of thumb". For example, there is a commonly used rule in the United States for annual forest improvement activities. Stands in a given operating unit are ranked on the basis of predetermined "average" rates of return that can be expected from various timber improvement activities in different categories of stands, which are defined in terms of site, stocking and age. The annual budget for stand treatment is then allocated to the ranked stands until the budget is depleted. The point to note here is that each stand is not subject to a complete appraisal each year before budget decisions are taken. Rather, rules of thumb are used for choosing projects, and these rules are based on detailed analyses done in the past of returns of various types of activity.

In contrast, a full-scale appraisal is much more frequently done for the "one time" type of decision, even though information on similar projects done in the past may be included in the appraisal.

It is widely believed that most decisions, including those for one-time projects, are based primarily on intuition and political expediency. Though we do not deny that a number of political factors enter into decisions of these types, we wish to stress that factual information and formal evaluations do contribute to decisions. Project appraisals, as we have seen, are not synonymous with decision-making; at best, they provide a useful basis for eliminating some alternatives and identifying more acceptable ones, which means that choices will still be affected by a number of intuitive judgements and subjective factors that are not easily included in a systematic appraisal. The essential point is that full-scale analyses or appraisals are needed when the cost of a penalty for a wrong decision is high. This, in turn, depends on the relative importance of the decision being taken.

Choices are also affected by the degree of uncertainty that surrounds a project. A good appraisal will never assume certainty for the estimated value of parameters. On the contrary, a good appraisal team will try to do sensitivity analyses, which consist in analysing the expected changes in project effects due to arbitrary changes in a given set of parameter values. Indeed, testing is usually concentrated on those parameters that are the most uncertain and that produce the biggest changes in project effects. As sensitivity analyses can provide the decision-maker with extremely useful information, they should be used to cover all types of project effects, and not just economic ones.

B. How Project Appraisals are Organised

The amount of work involved in a project appraisal can vary from man-days to man-years, depending upon the project and its purpose. The first rough appraisal can generally be done by the project initiator in a short time. However, the final thorough-going appraisal of the handful of promising project alternatives that remain after an extended period of design, appraisal and redesign, could keep a professional appraisal team busy for several months, if not longer. During this time, it is important that the appraisal team and the decision-makers remain in close contact, so that the appraisal will constantly focus on the most relevant effects and will continue to move in the right direction.

For the assessment of economic effects, a fairly long tradition exists with relatively stable, well-defined rules. On the other hand, the description and evaluation of environmental and social effects are much less advanced, even though some progress has been made over the last decade. Despite the fact that environmental and social effects cannot as yet be fitted into a conventional quantitative economic appraisal scheme, there has been growing agreement on the usefulness of including them in project appraisals.

At the outset we would like to stress that the extent to which a project effect is forced into a monetary framework is entirely a matter of policy. It is up to the public institutions involved. We see neither inherent good nor bad in attempting to develop monetary measures of value (shadow prices) for goods and services that may not have a market price, such as outdoor recreation, water and so forth. Basically, what is needed in project appraisal is a clear, adequate description of each effect, whether it be put in monetary or physical terms, or both.

For example, if a decision has been made to plant oaks that will be ready for harvest in about 200 years, then all that need be said is that the decision-makers judge that the value of the oaks will be high enough to justify the costs. A precise monetary value for the oaks is quite irrelevant. However, if a decision has not yet been taken, then the proper role of the appraisal team is to develop various measures, including economic ones, of the benefits to be derived from planting the oaks, as well as the costs. The benefits include not only the estimated value of the harvest but also such intermediate benefits during the growth period as protection, aesthetic pleasures, and the like. As for the costs, they are both the immediate investment and subsequent maintenance.

We believe that the best way to ensure a proper description and evaluation of non-economic effects is to include environmental and social science experts in the appraisal team, or at least to secure access to such expertise. A complementary strategy is to tap the knowledge of the local population.

A full-fledged appraisal team should therefore be comprised of foresters with sufficient knowledge of forestry and technological possibilities so they can design the project in detail; economists capable of conducting the necessary economic and financial assessment; and lastly environmental and social analysts who can not only identify the probable effects of the project in their respective fields of expertise but also pinpoint the public's view on the desirability of these effects. They should also be capable of suggesting changes in project design so as to create less problems and enhance benefits. Though this means expanding the capabilities of the appraisal team, we are by no means arguing in favour of impractically large teams. We simply point to the type of expertise that should be available during the appraisal process, either on a continuous basis or as needed.

But what effects should be included? An appraisal is at best harmless and at worst a waste of time and money if it offers information that is irrelevant either to the people involved in the decision and/or those affected by the project.

In very general terms, on can say that the goal of project planning is to identify a technically feasible solution to a problem; it should bring an acceptable economic return to society, be adapted to the institutional and managerial context, have acceptable social and environmental impacts and should be able to be financed with available resources.

Many trade-offs must be made to even come near this goal. Most forestry projects end up being compromises that move the nation ahead in some ways, while avoiding serious conflict with the other objectives. The compromise solution is reached through a process of give and take, in which alternatives are examined and adjustments made until an acceptable design is arrived at. To make this process effective, it is important that the appraisal team know what importance the decision-makers attach to the various objectives; in this way, they can concentrate their effort on analysing a range of alternatives that provides variations while remaining relevant.

In some instances, the decision-maker has well identified objectives, which makes it easier to organise the appraisal process. But in other cases, the decision-maker may have an intuitive sense of his objectives but will be unable to express them clearly. It is up to the analyst to help him and not to substitute himself by recommending an objective.

There are also cases where several decision-makers are involved in a single decision-making process. They may have different and sometimes conflicting objectives that are more or less clearly expressed. Here, a major task of the analyst is to build a common appraisal language. He will have to take into account the various objectives and help the decision-makers negotiate objectives so that a workable consensus can be achieved. The point here is that a well organised appraisal can help reduce the areas of conflict by providing objective information on trade-offs and the interactions of likely project effects.

Lastly, there can be cases where the decision-makers themselves are not clearly identified. Many persons are involved in the complex decision-making process but no one in particular can be considered as "the" decision-maker. In such a situation, the analyst's job is difficult indeed; he must not only appraise the project but also analyse the decision-making process. Otherwise, his work will be a purely academic exercise.

C. The Goal of this Handbook

We would like this study to serve as the basis for the development of manuals on forestry investment project appraisal at the country level. Such manuals should take into account the conditions, constraints and analytical capabilities of the country in question. Among other things, these manuals should reflect the organisational or institutional situation in the country and the economic and social objectives of the nation within which forestry projects will be designed, evaluated and chosen.

Here, we have tried to present the conceptual framework for such project appraisal. In the following chapters, we shall try to do the following:

- describe in broad terms the sequence of steps to be taken in a practical, useful appraisal process;
- indicate the very broad spectrum of effects that may be relevant in different projects;
- tentatively suggest how these various effects can be identified, described and evaluated.

We have therefore concentrated on the appraisal process and on the types of information that can be relevant. We discuss neither techniques for generating information nor specific sources of data or means for collecting and processing data. The goal has been to produce a practical framework for the multitude of small and large forestry investment projects that will be initiated in the OECD countries in the coming decades. The study is aimed at decision-makers in the forestry sector, for it is they who will most likely initiate project appraisals. It is also intended for the heads of appraisal teams, for it is they who will be responsible for carrying out the appraisal and completing the report.

Lastly, although this handbook could be applied to all types of projects, we have chosen to identify and describe those effects that are of interest to society at large: to individuals, to social groups, to institutions and even to future generations if there are long-term project effects.

The appraisal process that will be examined in the following chapters is already being used in a number of countries. True, the level of detail may differ or one step may be emphasized more than another. But this depends upon a variety of factors, such as the importance of the decision (the size of the project in relation to the size of the budget), the nature of the decision (is it a new one-time project or a routine activity?), and the political sensitivity of the project.

Why then, it may be asked, undertake a study on something that is already in use? There are a number of reasons. First, the appraisal process may be common practice but it is not necessarily understood or recognised as such. A clear specification could help to ensure more systematic application, especially when dealing with controversial projects. In such cases, a clear specification of procedures is essential to justify decisions, reduce conflicts and increase the accountability of public agencies.

A second reason is that while intuition and experience can lead seasoned administrators through the steps in the appraisal process, they will be saved a good deal of work once the process is set forth explicitly as a procedure involving systematic, well defined steps. Moreover, it not infrequently happens that the appraisal process is only partially understood and put into practice, making for incomplete appraisals: not all points of view are presented nor are all project effects.

Lastly, and this is perhaps the strongest justification for such a study, describing a proven informal process in a systematised, formal fashion makes it much easier to apprehend; this means that it can be more widely adopted by countries and agencies that do not presently use it. The basic point is that we are not proposing a new appraisal system for forestry projects; rather, we are synthesizing in an orderly fashion the best of what is already known and routinely practised in many countries as part of the decision-making process in the forestry sector.

The fact that we have combined the best elements from different systems and countries may make the process look just a little too perfect. However, we believe that every procedure proposed here is realistic. With a little imagination and, initially at least, a lot of hard work, the system is eminently practical. Though first attempts may not be perfect and may even be downright discouraging, we feel that eventually results will be worth the effort in terms of improved effectiveness of public administration procedures in the forest-based sector.

Long hours were spent debating how far we should move ahead from the typical approach today towards an "ideal" approach. What appears in the following pages is far from what we would consider as an ideal approach to project appraisal in a perfect world. Ideally, if there were no limitations on data, time, money and methodology, we would propose an overall systems approach, in which all effects would be described, and all trade-offs between effects would be quantified in a single, all-inclusive computer model. The best alternative would flow unequivocably and directly from simulations based on this model. Unfortunately, we do not live in such a world. Given this, we believe that what we propose here is the best practical approach. It carries the decision process one step closer to the ideal, even though the latter is still a long way off.

NOTE

1. We use the word "evaluation" to denote the process of comparing one type of effects with chosen criteria. The process of aggregating the different effects into a single, overall figure of "project desirability" requires further information from the decision-maker, for instance, the weights he gives to different objectives. The analyst cannot do this job for the decision-maker, but he may help the latter by giving him a clear overview of the different effects as well as information about how others judge the project desirability.

Chapter II

THE APPRAISAL PROCESS

A. The Main Stages in Project Appraisal

As we saw in the preceding chapter, the appraisal process is an iterative procedure comprised of project design, appraisal, redesign and further appraisal. Throughout the process, an increasing amount of effort is spent on assessing a declining number of design alternatives. The final stage entails a detailed description of the effects of those few alternative designs that have emerged as the most suitable, including a thorough description of the likely consequences of no action, known as the "Without" alternative.

The number of stages involved in an appraisal will depend upon the situation. With routine types of activity, one can sometimes move directly from the project idea to a final appraisal of two or a few limited alternatives that have been identified thanks to previous experience. In cases where experience and information are lacking, it will be necessary to go through successive stages of elimination and redesign before a detailed appraisal of the remaining alternatives can be made.

In a normal appraisal process, there are three main stages, during each of which the analyst will go through the three steps of identification, description and evaluation of the project effects, including economic/financial, environmental and social effects. (See Figure 2.1 and Table 2.1 for the three stages.)

It should be stressed that design and assessment must be closely interrelated from the earliest stage in project planning. If this is not done, and if project designers and appraisers do not interact as early as possible, then the appraisal will most likely be nothing more than window dressing, a kind of ritual that is followed for political reasons. Say, for example, that environmental concerns have not been incorporated at the outset; whatever negative environmental effects that turn up later will probably be treated only insofar as the law requires action or that these effects will translate into negative political effects. Moreoever, resistance to change builds up as the planning process moves ahead. This holds true for all types of effects: the more advanced the planning process, the more difficult it is to get project planners to change the design in response to the negative effects turned up in the assessment.

The implication is obvious: it is best not to take the first stage of the assessment process lightly. One should resist the common temptation to put off changes till a later stage when better information will be available. The first-stage assessment is probably more important than the final one, for the latter tends to involve only incremental adjustments to the work already accomplished.

Figure 2.1. STAGES IN THE PROJECT PLANNING PROCESS*

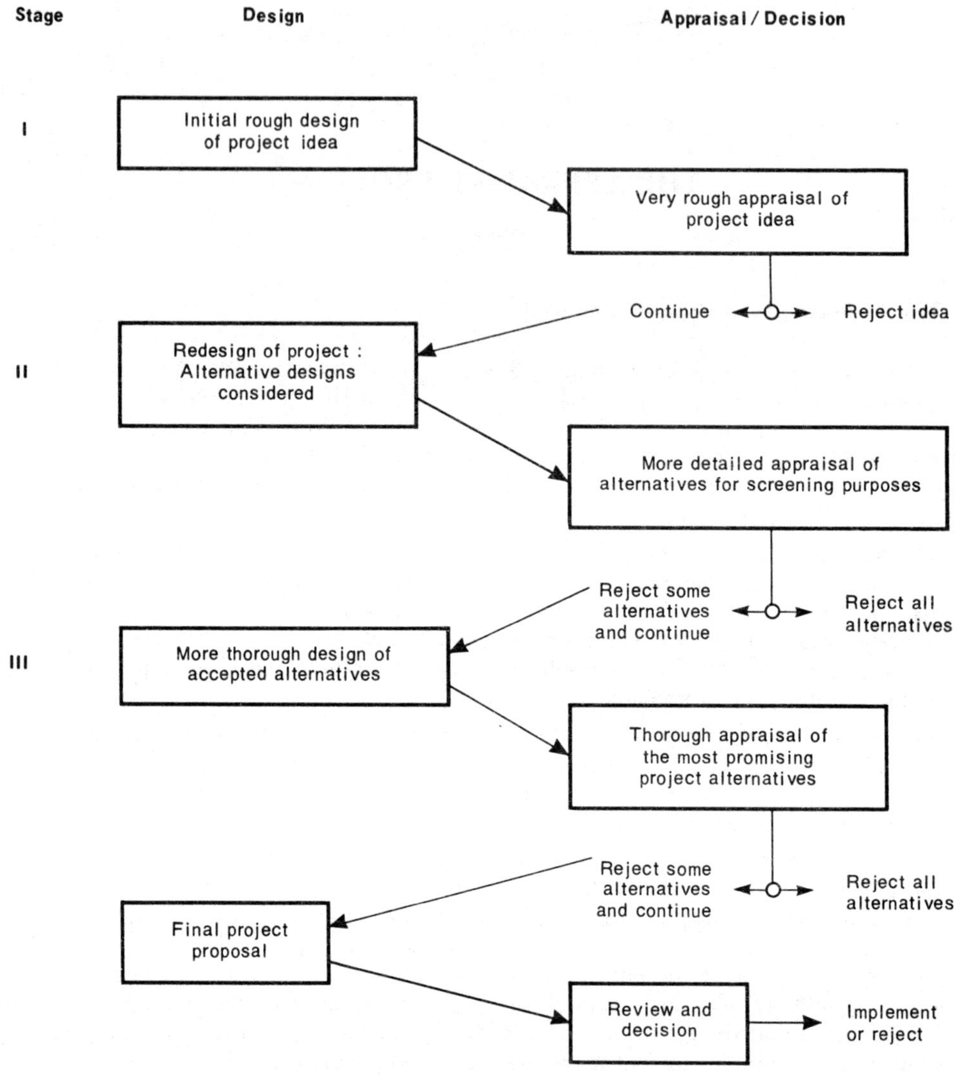

* Depending on the project situation, there may be fewer or more iterative stages.

Table 2.1. DESCRIPTION OF APPRAISAL STAGES

Stage I. ***Rough Appraisal of the Project Idea***

- Make tentative calculation of the economic effects of the "most obvious" project alternative and the "without" alternative;
- Make quick assessment of financial, administrative, and political feasibility;
- Attempt to detect adverse environmental effects (i.e., long-term and system effects), social effects and effects on different groups concerned (i.e. distributional effects);
- Consider means to mitigate negative effects.

Outcome: recommandation on whether or not to continue with project idea.

Performed by: project initiator, using existing available information.

Stage II. ***More Detailed Appraisal for Screening Purposes***

Using the results of Stage I, the appraisers:

- Design several project alternatives that seem relevant in light of existing objectives and of the major problems arising in the environmental and social fields, as identified in Stage I;
- Acquire economic, financial, environmental and social expertise for the appraisal;
- Make calculations of the economic and financial effects of the alternatives, possibly improving upon existing forecasts and shadow prices;
- Identify and describe the major social effects, possibly with the help of a representative discussion group;
- Identify and describe the major environmental effects, particularly indirect and long-term effects;
- Sample public opinion on the project alternatives;
- Exclude alternatives that are not feasible for administrative or political reasons;
- Establishing rankings of the remaining alternatives (as seen by various groups), possibly using the representative discussion group.

Outcome: Identification of several promising project alternatives, elimination of alternatives with obvious flaws, decision on whether to continue the project.

Performed by: Appraisal team in collaboration with external expertise and possibly a representative discussion group.

Stage III. ***Thorough Appraisal of the Most Promising Project Alternatives***

Given the results of Stage II, the appraisal team:

- Redesigns the project alternatives in the light of results obtained in Stages I and II;
- Completes the detailed analyses of economic/financial, environmental and social effects, collecting new data where necessary and utilising the insights of a representative discussion group. (Level of detail depends on the time and budget available, on the purpose of the appraisal and the nature of the project);
- Completes an appraisal report on the most promising alternatives, in the form of a scenario of likely developments over time for each alternative, including the "without" alternative;
- Ranks the most promising alternatives as seen by various groups, in collaboration with the representative discussion group, and possibly with input from local hearings;
- Prepares summary presentations of the scenarios and the ranking results for the decision-makers, public groups, financial institutions and other authorities.

Outcome: The necessary basis for choosing between project alternatives or terminating the project.

Performed by: Appraisal team, in collaboration with the discussion group and whatever expertise is available given budgetary and time constraints.

B. Some Basic Principles of Project Appraisal

The following general considerations hold good for all stages of the appraisal process.

1. Specification of a Project Alternative

In order to describe the effects of a project, one must first describe, or specify, the project itself. For example, in order to describe the effects of a reforestation project for marginal farm land, one must first know how much land will be involved, what equipment and labour will be needed, what jobs will have to be completed, and at what times. In short, one needs to know about the physical inputs.

Next, one must know about the physical output for example, the expected amount ot timber and pulp wood; the reduction in river siltation; the social changes in the use of the land; and again, when the outputs will materialise.

Good specification of project inputs and outputs are crucial for the economic and financial assessment for they are, after all, calculated as sums and as differences of inputs and outputs multiplied by appropriate prices or value estimates. Environmental and social assessments are equally dependent upon good project specification, but in a different way. The changes in inputs and outputs provide basic information that is used to generate ideas on likely environmental and social effects.

It is usually the lack of basic information on inputs and outputs that causes problems, and not the calculation of the effects of the project. This point cannot be stressed too much. It is not possible to describe project effects of whatever type without a proper specification of the physical inputs and outputs. Sometimes the analyst is blamed for "inadequate analysis" when in point of fact, it is the lack of input and output information that is at the root of the problem. And this is the job of the technical experts involved in the project design.

Lastly, the physical input and output information should be presented in such a way as to make it possible to compute the difference with and without the project alternative under consideration.

2. Selection of Project Alternatives

One of the basic principles of good project appraisal is that at least two alternatives be considered, namely the "with" and the "without" alternatives. At each stage and for every alternative, one must ask oneself: "How will the situation differ with and without the proposed alternative?" As the following examples show, this is definitely not the same as comparing the situation "before" and "after" the alternative is introduced.

Let us suppose that the project involves retimbering land left unfarmed owing to rural exodus. If the project is not carried out, the land will undergo a naturel process of change. (The sequence of plant combinations will vary according to local ecological conditions.) First, there will be fallow land, followed by heath, which will gradually be covered by trees. These changes may take several decades or even centuries, and will lead to a climatic pattern of woodland. An active retimbering project can therefore be seen as an attempt to outpace this naturel process of change, but at the risk of producing a forest that is less well adapted ecologically. In this particular example, the economic, social and environmental effects of retimbering should be compared with the effects of natural change. It is this last sequence of developments and not a static continuation of the current bare land that represents the "without" alternative.

Another example of how "with" and "without" alternatives are different from "before" and "after": let us take a retimbering project designed to protect and restore the soil. The project will also reduce erosion in a catchment area, thereby reducing the silting of impoundment reservoir downstream. In this example, the "without" alternative would have to consider how erosion and silting would develop under the effect of natural environmental factors and existing farming and grazing practices.

Apart from this basic minimum of "with" and "without" alternatives, there are usually several other possible designs with different effects. To identify the relevant alternatives, one can use either a systematic method or a heuristic method.

The systematic method, which involves identifying all imaginable alternatives, has a major drawback: it quickly generates an impossibly large number of alternatives. Obviously, a computerised appraisal model could take them all into account, but this is rarely what happens in reality. In practice, the analyst has to try to reduce the number of alternatives by screening them on the basis of simple criteria.

For instance, if the task is to select a suitable species for a reforestation project, some species may have to be rejected because of the quality of the soil. Similarly, afforestation techniques that require frequent clearings may have to be rejected because of the cost of labour, and so on. Here, the problem of identifying reasonably good alternatives is simplified by applying the principle of separability, i.e. by first choosing the "best" species and then, independently, the "best" afforestation method.

In the heuristic method, the initial basic alternative is gradually modified to bring about improvements. When a modification improves some dimensions without worsening others, the new alternative is obviously better than the original one. This method is used frequently, but there is a risk of overlooking an "optimum" alternative that the systematic method would have uncovered.

In practice, the best method for identifying alternatives lies somewhere between the two. Starting with one or several basic alternatives, the appraisal team can systematically ask itself which modifications could bring about additional attractive alternatives or could improve existing ones. One basic principle is to include all those alternatives that are advocated by one group or another and to exclude all those for which no advocates exist.

3. Selection of Project Effects

Here again, one can use a systematic or a heuristic approach. Checklists can prove very useful when trying to systematically think of all the effects that might be of relevance. In the heuristic approach, a dimension is introduced only when a particular problem arises that threatens to torpedo the project. Formerly, social and environmental effects used to be taken into account only when pressure groups demonstrated their opposition to the project or when pollution problems or other obvious disutilities came to the public's attention.

As we have seen, the systematic approach has the advantage of being complete, but is liable to involve prolonged, tiresome and often useless studies because some of the chosen effects turn out to be irrelevant. The heuristic approach is convenient but is liable to overlook important effects and therefore lead to poor appraisals. Regardless of which approach is chosen, an effect should be included if it corresponds to:

- a concern of the decision-maker(s) involved in the project;
- a concern of one or several of the social groups affected by the project;
- a change in the natural or social environment which is irreversible or which may pose difficult problems in the future;
- an effect that the analyst feels should be included.

Any effect that meets one or more of the above criteria should be taken into account even if there is no known method for measuring it. A mere description of the effect points up present ignorance and furthermore prompts the use of sensitivity analyses using different assumptions about the strength of the effect.

Another question is the number of effects to select. In order to focus the attention of decision-makers, it should not be too large. It is sometimes possible to make do with only one dimension. In France, for example, a method of ranking alternative sites exists for setting up fire protection areas. The method relies on the value of a single indicator, namely, the ratio of total damage avoided to cost. Total damage avoided is computed as a value index multiplied by the reduction of fire risk due to fire protection. The value index is set equal to the average age of the forest stand, corrected by a number of multipliers that take special characteristics into account.

There is a tendency to concentrate on effects for which sophisticated measurement methods exist, at the expense of effects for which only summary methods are available. This is a serious mistake. Ideally, the amount of effort spent on an effect should be in proportion to its importance. Therefore, if a particular effect is found to be especially important and no known method exists to measure it, then it would be wise to allocate a large part of the appraisal effort to developing a method and an estimate of the magnitude of the effect.

Finally, something should be said about the frequently raised issue of "double counting". In our opinion, it is more of a myth than a reality. The only way double counting can crop up is when someone deliberately whishes to confuse matters, perhaps to gain advantage for one type of project over another. It is not double counting of employment benefits if the analyst includes both employment gains and income gains for those employed. The two measures are not combined, and they each provide useful information to the decision-maker. For example, if it is believed that there is a certain element of human dignity involved in holding down a job and that lowering unemployment can help to reduce crime or some other undesirable social effect, then both the income and employment measures consitute useful information.

Take, for example, a situation where there is a relationship between employment and crime. Two projects exists: both produce an additional $100 000 of income, but one creates 200 jobs while the other creates only 100 jobs. Both projects have the same monetary effects but quite different social effects. Consequently both pieces of information – the additional number of people employed and the additional earnings – must be included, for they are very important for the decision-maker.

In a similar vein, let us assume that a forest exploitation project will increase erosion on a watershed, thereby increasing siltation in a river. The direct physical effect of the project is to increase erosion by X tons per hectare per year and to increase the sediment load of the river by some quantified amount. In terms of the social assessment, the effect may be to increase local dissatisfaction because of deteriorating aesthetic conditions. In terms of the environmental assessment, there may be more than just the negative effect of direct soil deterioration which, in some case, can be considered as almost irreversible. The analyst may identify negative effects on the fish population or other instability effects. There are thus three useful measures of the same direct physical effect, that is, increasing erosion. It is not double counting to include all of them in the appraisal. On the contrary, they should all be included to make the appraisal complete.

4. *Treatment of Long-term Effects*

Till now we have said very little about the time dimension, although it plays a vital role in most forestry projects. For instance, a management plan for an oak forest may be designed to cover only 20 to 30 years, but some of the trees may not be felled for over 200 years. Even fast-growing varieties may not be harvested till 20 to 40 years after planting. This is a very long time compared with the amortisation of investment or equipment in other sectors of the economy.

The normal way to allow for time in most project appraisals is to use a discount rate. This is suitable for dealing with short-and medium-term effects (say, 10 to 20 years). But when dealing with forestry investment projects where there is a significant time lag between investment and marketable output, other methods may be relevant to highlight the long-term consequences of the project alternatives. They can be based on an explicit description of the long-term impact of a project, and particularly non-economic effects.

When it is possible to use the discount rate (for example, with a quantified dimension), it is reasonable to present a discounted measure that sums up the whole of the time dimension in a single figure. However, even when the validity of a discount rate is beyond dispute, it may be desirable to present the decision-maker with a timetable or with an explicit description of the pattern of development over time. This can take the form of tables or graphs.

In order to compare alternatives for a project that has a number of effects over time and along different dimensions, we recommend establishing an impact table that shows how the different alternatives affect the most important dimensions over time. To make sure that the long-term effects are properly emphasized, a column can be included showing the estimated impact on the very long-term value of the project area. This is useful even if only summary methods are available for measuring the effects. The aim, after all, is not to obtain a set of highly detailed pictures but to make the decision-maker aware of the most important effects and of how they can vary according to the alternative.

5. Treatment of Uncertainty

The information used in project appraisals is approximate: it is, of course, based on the best data avalable, but these data may not be very good. *A fortiori*, the same holds true for the monetary, physical and qualitative effects of a project in the future. Yet somehow one must come grips with the problems created by having to use imprecise data or future estimates. In rare instances, various forms of probability analysis give some idea of the range of values that are likely at different levels of probability. However, in most situations, probabilities cannot be calculated; that is, one is dealing with genuine uncertainty.

The most practical way to deal with uncertainty is to carry out a sensitivity analysis for the project. This entails testing the sensitivity of chosen measures of project performance, in the alternative assumptions.

The first step is to identify the main sources of uncertainty. This gives the appraiser and the project planners a first approximation of the parameters to be tested in the sensitivity analysis. For each of these critical sources of uncertainty, the appraisal team estimates a reasonable range of values (monetary or non-monetary). It then tries to estimate what effect changes in assumptions about such values will have on the outcome of the project.

For example, in the case of economic effects, the team may recalculate the internal rate of return for the project, using different assumptions about the prices of inputs and outputs. For social impacts, the team may look at the likely effects if the assumption concerning public acceptance of the project is changed. The initial information used may be in the form of a qualitative judgement. The sensitivity analysis also involves a qualitative judgement. What, for example, is the effect if less people than initially assumed support the project? Or, say, the environmental analysis leads to the conclusion that there will be little effect on erosion.

The number of parameter-value changes tested depends upon the budget and the time available, the perceived importance of the parameters and the possible ranges in values (monetary and non-monetary) that are judged to be reasonable. There is no pat formula to determine the number of sensitivity analyses that are needed. This is up to the judgement of the project planners and appraisers. At the minimum, several different scenarios should be tested. When a reasonable change in parameter value is critical in terms of some dimension of project performance, this may indicate that the planning team should study that parameter and its possible values and impacts more closely.

The sensitivity analysis may not do away with uncertainty, but it can give the decision-maker a clearer idea about the risks involved in the different alternatives and allow him to make as well-informed a decision as possible.

6. *Comparison of Project Alternatives*

A project alternative will usually have many different types of effects; it can, for example, affect employment, timber output, recreation, and aesthetic enjoyment. Some of these may be very positive effects, while others will not score nearly as high. If there is no common unit to measure all the effects, it becomes very difficult to compare alternatives and pick the "best" one. The decision-maker's task is obviously simplified if the appraisers can come up with a single index of project desirability.

Some analysts try to present a single index of project desirability by putting money values on all the project effects and by summing up the picture as one measure (usually discounted) of net present value or as a cost-benefit ratio. This can be useful when the effects are primarily of an economic/financial nature, but if there are a number of intangible effects that are difficult to put prices on, which is the case of many forestry projects, we believe it is better to present the decision-maker with a description of the effect, rather than trying to fit it into some overall measure of project desirability.

There are basically two ways in which project effects can be combined into a single index of project desirability: "shadow pricing" and "weighting". Both are basically the same, since shadow pricing is nothing more than a systematic way of putting monetary weights on effects in order to reflect benefits to society in monetary terms.

Let us take a hypothetical multipurpose project that is designed to protect certain threatened animal species in a given area while allowing for a moderate amount of lumbering and some access to the public. Let us suppose that there are three alternatives, each of which puts emphasis on one of these more than on the other two. (See Table 2.2 for the main effects). How can an index of desirability be developed for the four alternatives (A, B, C, and "Without" the project)?

Table 2.2. A HYPOTHETICAL CASE:
EFFECTS OF PROJECT ALTERNATIVES

Effects	Project alternative	"Without" project	A	B	C
I	Cost of establishing the project ($)	0	10	15	5
II	Operating cost ($ per year)	0	2	3	1
III	Timber production (m^3 per hectare per year)	0	3	1.5	1
IV	Number of visits (1000 persons/year)	10	30	100	20
V	Quality of landscape (index)	4	3	4	5
VI	Number of species protected	10	8	5	15

With the shadow pricing approach, which attempts to give a single monetary measure of project desirability, the first step is to combine effects I and II into a single monetary measure. This mean choosing a rate of discount, for example, 10 percent a year, and converting annual operating costs into an equivalent present value.

Effect III, production, can also be incorporated into monetary effects, provided that a shadow price is determined for the timber produced. For the short term, it could be that of the present market price for wood, but this presents a problem since market prices fluctuate depending upon economic conditions. And, in the long term, the price of wood is fundamentally uncertain.

As for the number of visitors (IV), one could conceivably incorporate it into monetary effects by putting a shadow value on each visit, although there is no indisputable method for calculating such a value. Similarly, trying to put a shadow price on each animal species protected (VI), or on each point on a scale of landscape values (V) is a debatable procedure, though it can be and sometimes is done. In short, the choice of values is more or less subjective.

The same holds true for the weighting method, which consists in giving each project effect a non-monetary weight and then adding up all the points. This method presents the advantage of being easy to describe and apply, but it should be used with great care for several reasons.

First, using a weighted total is in a sense the same as using arbitrary shadow prices. If a weight "p" is given to Effects III, it is easy to calculate the shadow price for a cubic meter of timber which corresponds to this weight. Likewise, a weight of "q" for Effects VI could be translated into a shadow price for protecting the species. The conclusion is that weighting gives no better results than shadow prices.

Second, one must be explicit about the units used for measuring each type of effect. To say, for example, that Effects V (quality of landscape) has a weight "r" is meaningless unless it is specified that it is being rated on a scale of 0 to 5. If the scale were from 0 to 20, its weight would obviously be less important.

The third and last disadvantage of weighted totals is that all effects are assumed to be independent for purposes of the decision-maker's preferences, that is, that there is no interaction between the different effects.

Consequently, though it would seem that weighted totals are easy to use, they are in fact very difficult to handle if one is to avoid misleading conclusions. Determining appropriate weights is essentially part of decision-making unless, of course, an acceptable, agreed-upon weighting system exists. Once the weights have been agreed upon, it is a simple matter to attach them to the measures of project effects and to obtain an overall index of project desirability.

A completely different method for choosing from among different alternatives is known as the "constraints approach". Using the same example, let us assume that the decision-maker wants to see that the following constraints are taken into consideration: the cost of establishing the project should not exceed $12; the operating cost should not exceed $2.5/year; production should reach at least 2 cubic meters per hectare per year; the number of visits should be at least twice as big as in the "without" alternative; and the value of the landscape should not be reduced and the ecological value should increase.

None of the alternatives proposed in Table 2.2 takes all of these constraints into account. They should therefore be rejected and new alternatives should be developed. Let us call the new alternatives D and E and assume that they have the effects shown in Table 2.3.

D and E comply with the decision-maker's constraints, and it is now up to him to choose between these two very similar alternatives: either the one that favours productionD) or the

Table 2.3. A HYPOTHETICAL CASE:
ADDITIONAL PROJECT ALTERNATIVES

Effect	Alternatives	D	E
I	Cost of establishing project ($)	12	12
II	Operating costs ($ per year)	2.5	2.5
III	Timber production (m^3 per hectare per year)	2.5	2.5
IV	Number of visits (1000 persons/year)	30	20
V	Quality of landscape (index)	4	5
VI	Number of species protected	11	13

one that favours protection E). Compared with Table 2.2, his choice is narrowed down and should be easier to make. The difficult questions have, in fact, been transferred to the choice of constraints, but this can be easier for the decision-maker to deal with than a set of weights.

In actual practice, the weighted totals or the shadow prices approach is often combined with the constraints approach. In many situations, it may be more convenient to express in terms of shadow prices or weights those effects that can be measured relatively accurately and reliably, and to express in terms of minimum constraints those effects that are less easily measured. Such minimum constraints (especially environmental and social ones) often prove useful during the preliminary screening of alternatives, because they exclude from the subsequent appraisal those that rank too low in any one type of effect.

C. Presentation of the Project Appraisal

Drawing up a good project appraisal report is by no means an easy task. Probably the biggest difficulty lies in extracting the few essential points from the mass of information that has been collected. The appraisal team has to be able to simplify it so that the decision-maker can understand, retain and use the results. Simplicity, in turn, requires that the appraisal report be short, that it cover reasonably few alternatives and that it be readable.

The appraisal report is, of course, drawn from the extensive and, to the layman, obtruse working documents of the appraisal team. As these documents are normally not studied by anyone other than, say, a handful of outside experts, it is superfluous to spend time editing them. However, it is essential that they be as complete as possible, describing everything that has been done.

It is normally a good idea to divide the presentation into two parts, the first part describing the individual project alternatives and their effects, and the second part comparing the effects of the various alternatives. A presentation of the development over time should also be given for each alternative, preferably starting with the "without" alternative.

Each alternative should be presented in the form of a written scenario, possibly illustrated with graphs showing the development over time for the most important variables.

Figure 2.2. HOW TO SHOW A QUANTIFIED VARIABLE OVER TIME

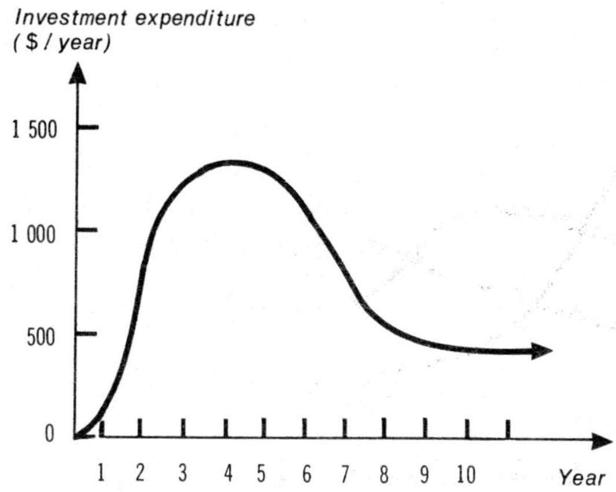

DISCOUNT RATE (%/YEAR)	NET PRESENT VALUE OF INVESTMENT STREAM ($)
7	800
10	650

Figure 2.3. HOW TO PRESENT A SPLIT QUANTIFIED VARIABLE

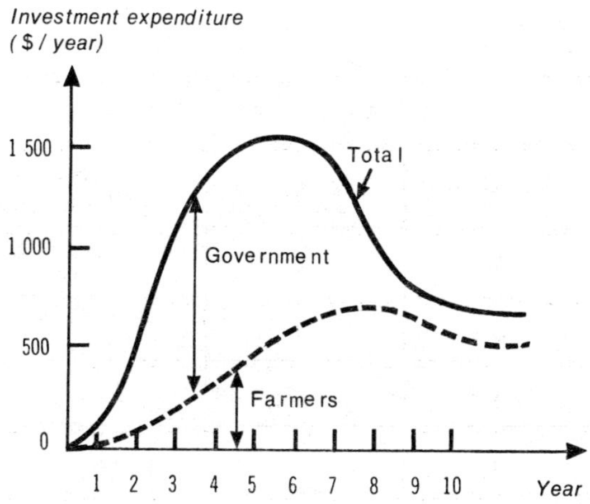

DISCOUNT RATE (%/YEAR)	NET PRESENT VALUE	
	GOVERNMENT	FARMERS
7	600 $	200 $
10	500 $	150 $

Figure 2.4. A HYPOTHETICAL CASE: COMPARISON OF PROJECT ALTERNATIVES

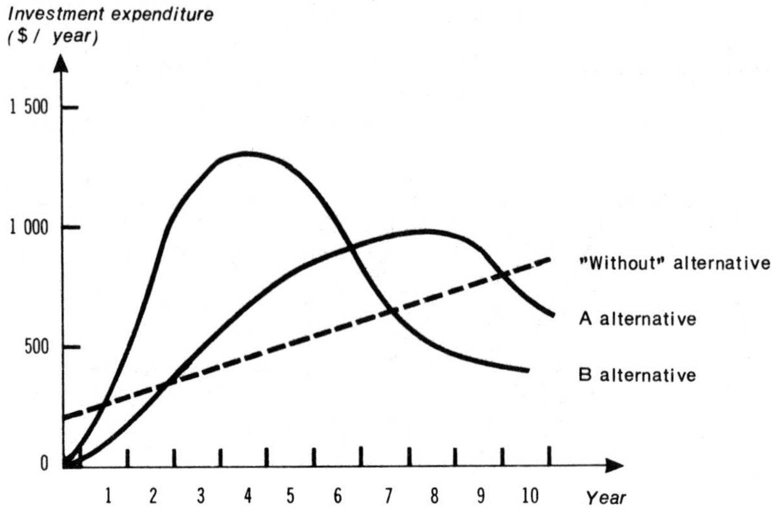

Figure 2.5. A HYPOTHETICAL CASE: RANKING OF FOUR ALTERNATIVES BY SOCIAL GROUPS

ALTERNATIVE	SOCIAL GROUP		
	ALL VOTERS	MEN OVER 50	LOW-INCOME GROUPS
A	1	2	2
B	2	3	1
C	3	4	4
"Without"	4	1	3

In Figure 2.2, for example, the pattern of investment expenditure over time is shown. It might also be advisable to compute summary measures, like the net present value for different discount rates, as shown on the right-hand side of Figure 2.2. If it would be useful to describe the distribution of an effect among social groups, this can be done in the written scenario as well as in a graph. In Figure 2.3, for example, the total investment expenditure has been split between the farmers and the government.

Obviously, tables can be used but often the numbers can look much too precise, given the uncertainty of the data base. A line should be completed in the impact table for each of the project alternatives.

Every project appraisal should include an "executive summary", a one-to-four-page statement that summarizes the objectives, the alternatives considered, the main results of the appraisal, the results of the sensitivity analysis and the conclusions and recommendations.

Once the effects of each alternative have been described in a way that gives the reader a good understanding of the time path, it is important to try and make comparisons of the alternatives as clear as possible. Talks between the appraisal team and the decision-maker can highlight the principal differences between the alternatives.

The basic tool for such comparisons is the impact table. For quantified variables, comparative plots are instructive. Figure 2.4 shows the investment paths for the A, B and "Without" alternatives in a hypothetical project. Tables are also a possibility, since some users find them easier to use than graphs.

As a final aid, it may be useful to present a ranking of the project alternatives done by different social groups, as shown in Figure 2.5.

The sensitivity of appraisal results to variations in especially uncertain parameters must be presented; to not do so is inexcusable. True, such presentations can often be quite confusing, but this can be remedied somewhat by restricting the presentation to a few extreme cases, shown either as possible time developments (a shaded area in a graph), as a possible range of summary measures, or as an indication of the maximum shift in ranks when parameter values are varied.

We will now turn to the central concern of any appraisal: the identification, description and evaluation of project effects. For the sake of convenience, we have divided the discussion into three parts, one for social, one for environmental, and one for economic/financial effects.

Chapter III

SOCIAL EFFECTS

Forestry investment projects can often have important social effects, regardless of whether the project is to reforest, to increase harvesting or to build roads. As used here, a "social effect" is any impact on people that is not economic or environmental in character. A more technical definition restricts social effects to anything not covered by a shadow price in the economic analysis[1] or by an environmental impact.

Examples of social effects abound. A decision to invest in harvesting machines may deprive many forest workers of their jobs and force them to migrate. A programme to afforest a border county may increase the national presence in, and thereby national control over, the area. The use of insecticides may produce adverse genetic effects. The introduction of vibration-free chain saws may improve working conditions for forest workers. The establishment of export-oriented timber production in a local community may make the level of community employment increasingly vulnerable to fluctuations in the international markets. A shift from floating timber to road transport may destroy the cultural tradition attached to floating activities. These examples – increased migration, increased national control, adverse genetic effects, improved working conditions, increased vulnerability, and destruction of cultural traditions – illustrate the broad range of possible social effects.

With increasing affluence, social effects have been given increased attention. Today, in rich OECD countries, forestry investment projects may be stopped, or at least significantly modified, because of the unacceptable social effects of the initial design.

Assessing social effects is especially difficult given their exceedingly broad range and diffuse nature. It is impossible to know ahead of time what intricate chain of cause-and-effect may be triggered as a result of a forestry investment project. The situation resembles that of environmental appraisal ten years ago, before information on the most common environmental impacts was accumulated and made available both in the literature and through courses in environmental science.

In addition to the problem of identifying social effects, there is that of description, particularly of their development over time. In all but a few cases, measurement is out of the question, so it is impossible to put the various social effects on a common denominator. At best, one can try to rank the project alternatives in light of the social effects that have been identified. Judgements may be made somewhat more "objective" by consulting as broad a cross-section of the population as possible.

In short, assessment of social effects is primarily an art and not a science. Innovative efforts, inspired by the accumulated wisdom of the local population, and using the few available fragments of data, may reveal ahead of time major social consequences. On the other hand, uninspired assessment work will rarely yield results of any importance, although it can sometimes lead to minor alterations in the project design.

A. Questions Decision-makers Need Answers to

In spite of these difficulties, decision-makers need answers to three broad questions.

1. What social effects will occur when the project is implemented?

   The problem here is to identify social effects. Some effects, such as employment or rural development, will be obvious while others, such as loss of traditional skills or shifts in the power balance between local groups, may be harder to discover. Obviously, immediate effects will be more visible than delayed social effects, and tangible effects will draw more attention than intangible ones.

   In any case, the decision-maker will be greatly helped by a complete list of the most likely social effects of the project. Even if it proves impossible to determine the strength and the timing of an effect, simply being aware of its existence gives an increased sense of control and confidence in the decision process.

2. How strong will the social effects be at different points in times?

   In any case, it will not be possible to answer this question. Most social effects are difficult to measure quantitatively. One can therefore do little more than describe as carefully as possible the likely social effect and its development over time. However, any information that helps the decision-maker determine which social effects he can neglect and which he must take onto consideration will facilitate his task.

3. Who will be affected by the social effects?

   Social effects do not affect the population uniformly. Some groups may be badly hurt, while others may not notice the effect at all. Still others may actually benefit from the same effect, e.g. they might enjoy quieter surroundings after a major exodus of unemployed. Thus it is exceedingly important for the decision-maker to know who will be affected, and for the appraisal to disaggregate the population affected in a readily understandable way.

The task of the appraisal team is to give the best answers possible, given the limited time available, and to make the level of uncertanty explicit. The work should proceed in an incremental, iterative fashion. At each stage, there should be a critical review of the remaining project alternatives, possibly ending with exclusion of some alternatives and revision of the design of others.

In addition to answering the three questions above, the decision-maker may well want to know what public reactions are likely to be triggered if the project is implemented (political effects), and how the project design could be altered to reduce the most offensive social effects?

These questions are even harder to answer, and have to be left to the more spontaneous thinking of individuals in the appraisal team, among the decision-makers and in the public debate around the project.

B. Procedure for Answering the Questions

The assessment of social effects should proceed in three stages, in line with the process described in Chapter II. Each stage encompasses the three steps previously outlined: identification, description, and evaluation of social effects. Identification and description are done to help answer the three questions in the preceding section. In addition, an evaluation of the importance of the social effect is performed either by the appraisal team, by some representative group drawn from the affected population, or by a referendum. The results are presented to the decision-maker to improve his basis for choosing between project alternatives. In the rare situations where the social effect can be measured in a relatively indisputable and inexpensive manner, description can be replaced by measurement.

Obtaining public participation is vital, because the local population can be an extremely useful source of ideas about likely social effects. Under no circumstance should it go untapped. Local knowledge can help save construction time if taken into account in the design. Public participation can, moreover, ensure some adaptation of the project to local social and cultural conditions, and can secure political support for the project. Conflicts may be uncovered early enough to be minimised through negotiation and mutual education.

A "discussion group" of representatives of the various relevant interest groups can be the best basic tool for social appraisal of investment projects. It is probably the most practical technique at the moment, being both fast and inexpensive, particularly in conjunction with formal expertise. Our advice on how to organise the participatory process is a central element in our suggestions on how to complete the three stages of social appraisal.

C. Procedure During Each Stage of the Appraisal Process

What follows is a brief description of how the steps of identification, description and evaluation are carried out at each stage of the appraisal process. This is followed by an example that illustrates the process in more detail.

Stage I: A very rough social appraisal of the project idea

Identification

In the first rough social appraisal of the project, the task is primarily to identify social effects that might be sufficiently grave as to warrant termination of the project. The task has to be completed by the project initiator, who usually does not have much experience in social analysis. To a large extent, he will have to rely on common sense, and on inspiration from the three questions listed above. Checklists of common issues are also of great help. One such checklist is given in Table 3.1.

As a first step, the project initiator should try to decide whether any of the effects in the list could be of decisive importance. The task will be much simpler if undertaken in collaboration with someone who knows the affected area.

Description

In the rough social appraisal, there is little need for an elaborate description of the social effects, whom they affect, their time development, etc. All that is needed is to point out those that are crucial, if such have been identified.

Evaluation

The evaluation process consists of judging each social effect as acceptable/unacceptable. This is the job of the project initiator, but he should try to put himself in the place of the public concerned.

Stage I: Summary

Identify effects using checklists and common sense. Judge the effects as acceptable/unacceptable to the affected population

Table 3.1. CHECKLIST FOR COMMON SOCIAL EFFECTS OF FORESTRY INVESTMENT PROJECTS

Effects on employment

Does the project affect:
- *– the number of jobs?
- *– the location of jobs?
- *– the type of jobs?
- *– educational requirements?
- *– job stability?

Effects on working conditions

Does the project increase/decrease:
- – the physical strain on the employed?
- – the psychic stress on the employed?
- *– the on-the-job accident rate?
- – the job status of the employed?

Effects on Population Structure

Does the project affect:
- *– the rate of in- and out-migration?
- *– the composition of migration flows?
- – the destination of migration flows?
- *– the age structure of the population?
- – crowding in local centres?

Effects on Community Structure

Does the project:
- – have strong forward and backward linkages (multiplier effects)? Locally?
- – compete with existing business? Locally?
- – concentrate or diffuse decision power and participation?
- – imply major discontinuities in land use?
- – imply in- or out-migration of businesses and other institutions?
- – change the level of group conflict?

Effects on Social and Cultural Tradition

Does the project affect:
- – the "stage" or backdrop for some traditional activities?
- – an area of symbolic importance?
- – communication between social groups?
- – local values and attitudes?
- – maintenance of traditional skills?

Effects on vulnerability

Does the project affect:
- – local dependance on outsiders?
- – the sustainability of the current way of life?
- – the welfare of future generations in the area?
- – the regional heritage?

Effects on Political Independance

Does the project affect:
- *– the balance-of-payments?
- – national control over areas?
- – dependence on foreign technology?

Effects on Individual Quality of Life

Does the project affect:
- – the extent of public participation?
- – access to opportunities?
- – individual health?
- – access to education?
- – local communications?

* Indicates the effects that are easy to measure.

Stage II: More Detailed Social Appraisal for Screening Purposes

Identification

In the second stage of the appraisal, there is both the time and the opportunity to get the opinion of experienced social analysts (in or outside the appraisal team), as well as of the public. The usefulness of public participation cannot be stressed too much. By helping to pinpoint probable effects on the community, local representatives often come up with suggestions for project redesign which can well lead to a project that is better adapted to local condition before it reaches an advanced stage. This is important because the greater the effort that goes into a project, the more difficult it is to change later. So it is best to examine desirable changes as early as possible.

To ensure that a specific point of view does not predominate in discussions, the appraisal team would do well to take the time to carry out a preliminary investigation to determine which are the major social groups in the community. It will then be possible to set up a broadly based group of representatives for discussions, including, for example, professional groups, different age groups, labour and community officials, members of the school board, and so on.

However, for practical reasons, it is preferable to limit the discussion group to ten members, especially if it is to meet frequently. In selecting these community representatives, the appraisal team should keep an eye out for imaginative extroverts who are not too verbose. A trained social analyst can be of help at this stage, but if one is not available, a strong dose of common sense will go a long way.

For smaller projects, it may be sufficient to call only one meeting of the discussion group or to conduct a public hearing during which private individuals and appointed representatives can exchange views. To make sure that no important social effect is overlooked, the appraisal team can use a checklist.

Description

Even in this second stage of social appraisal, there is little need for an elaborate description of the social effects that have been identified. First of all, a thorough description, not to mention a measurement, is so time-consuming as to be impractical. Moreover, most decision-makers are quite content with an oral explanation of the chain of cause-and-effect through which a social effect will probably arise.

Once the decision-maker is made aware of a likely effect, he or she will more often than not immediately form an opinion on its importance. And this opinion will be extremely difficult to shake, even when elaborate scientific measurements are put forward to the contrary. This may not be an ideal situation, but it is nonetheless a very common one and goes to show why an elaborate description is not warranted when screening a broad range of project alternatives.

Evaluation

The discussion group is a practical tool for evaluating the social effects that have been identified and described (possibly in a small report listing the project alternatives and their likely social effects). The group should not be asked to evaluate the effects, but rather to rank the project alternatives either by descending order of preference or classifying the alternatives acceptable and unacceptable. The appraisal team can try to achieve consensus on one ranking (hoping that it will reflect public opinion) or it can keep separate and report the rankings of the various members.

Using these rankings, and the results of the economic and environmental assessments, the appraisal team will then try to identify those few project alternatives that warrant thorough social appraisal.

Stage II: Summary
Establish a representative discussion group. Identify effects using the discussion group, experienced social analysts and checklists. Give verbal description of the identified effects and their causes. Let the discussion group rank the project alternatives according to perceived social desirability.

Stage III: Thorough Social Appraisal of the Most Promising Project Alternatives

Identification

At this third stage, the major social effects have most likely been identified. Even so, the appraisal team should remain open to ideas gleaned in local newspapers, in political debates or in academic centres, among others. A final list should be drawn up of the likely social effects of each of the few remaining project alternatives.

Description

It is now time for the most complete description of the social effects that imagination, data and time allow. This takes the form of a written scenario describing the social developments for each project alternative overtime. It should explain not only what is expected to happen, but also why. Each scenario should cover all the social dimensions that were found to be crucial in any one of the alternatives and should indicate how the effects will be distributed among the various social groups. Needless to say, the "without" alternative should be covered in the same manner.

Some social effects are easier to quantify than others (e.g. those marked * in Table 3.1), and it may prove helpful to illustrate the scenario with graphs showing the expected time development for these measures.

Prediction of social consequences is difficult. Formal modelling and correlation studies are now being used, but they do not make for accuracy greater than what can be obtained from a broadly composed discussion group. Modelling may, however, increase the internal consistency of a scenario by insuring that there is no contradiction, say, between the labour costs assumed in the economic appraisal and the assumed employment effects in the social appraisal. Simple charts mapping chains of cause-and-effect may be of help in clarifying how some social effects depend on the development of some easily observable variable. Proportionality factors (for instance between the incidence of alcoholism and the number of unemployed) can be estimated by the appraisal team and/or the discussion group.

The reliability of such proportionality factors can be increased by conducting a simple poll; for instance, by asking "How many people do you think will migrate if the local per capital income level falls 25 per cent below that of the neighbouring communities?" But even such a simplified form of social measurement may not be worth the effort; a local inhabitant would think a great deal before deciding to migrate but not necessarily before answering a question in a public opinion poll. However, there are certain circumstances that justify a full-blown social study, and that is when an entire project may depend on the acceptability of a particular social effect.

Evaluation

The best approach is to rank the project alternatives according to their social effects. The discussion group can do a first ranking; if they can agree on a single ranking, this can be very useful. If they cannot agree, individual rankings should be presented. Another possibility is to add up the individual scores and use this as a measure of "political acceptability".

To further increase the significance of the ranking, a public hearing can be organised, during which the affected population is allowed to react on the written scenarios and on the preliminary ranking by the discussion group. In extreme cases, a referendum could be held, and the affected population could participate in the ranking process.

A hearing should be held on any large project, to avoid unnecessary blunders in design and reduce political oppposition.

The final report to the decision-maker should contain the compact social scenarios for the few remaining project alternatives, including, of course, the "without" alternative. The report should also contain the ranking of these alternatives by various groups. The report should not necessarily select one of the alternatives as the "best" one. The choice is, after all, the task of the decision-maker. If asked, the appraisal team can present its own view on the best solution, but such subjective judgements should be kept quite separate from the assessment of project effects.

Stage III: Summary

List the major social effects in each of the few promising alternatives. Write compact social scenarios for each alternative, including the development over time in all dimensions of importance. Include time plots of easily quantified effects. Achieve a ranking of the alternatives, using a discussion group and, possibly, public hearings. Complete a report describing the scenarios and the various rankings.

D. A Sample Case: The Inner Valley Road Project

To increase its wood supply, a paper company in the capital of Forestland proposes to build a logging road into the inner reaches of an inaccessible valley. This would increase the sustainable profitable cut in the valley by 100 per cent and make truck transport possible rather than the current skidding of timber through the forests. It would also make possible the use of advanced harvesting machinery. Till now, felling and skidding have been done by tractor and chain saw by farmers living at the outer end of the valley. They make an acceptable income by working the land in summer and the forest in winter.

Calculations by company officials show that the use of modern harvesting equipment would more than double labour productivity per cubic meters harvested and transported. However, for the project to be profitable, the equipment would have to be used both summer and winter.

The company decides to have a very rough appraisal of the project done. Using a checklist, the local forester concludes that the following social effects would be significant: a fall in total employment (though volume would double, productivity would more than double, which implies a fall in the total number of man/hours worked each year; a change in working conditions, from many part-time to a few full-time jobs; from varied work to more industrialised production, and from risky to relatively safe work; a possibility of out-migration (those who would lose their forest job could no longer survive solely on their farm income); and a possibility of conflict (between those who got the few available forestry jobs and the others).

The very rough social appraisal concludes that the project as described is unacceptable because of the tension it would create in the Inner Valley Community. But further studies are recommended because alternative and much less socially disruptive project designs are easy to imagine. The economic and environmental assessments are much more promising, and a decision is made to proceed with the project idea.

Since it was found that the social effects are the most adverse, the company decides to employ an experienced social analyst on the appraisal team.

As a first step in Stage II of the appraisal, the social analyst decides to set up a discussion group with representatives of the paper company, the forest owners, the old as well as the young farmers (who are more technically alert and more likely to get the forestry jobs), the owners of the land under the projected road, the local community council, and a nearby college known to foster environmentalist attitudes. The discussion group meets, and in a series of long, aggressive meetings, participants finally agree that the list of social effects drawn up in Stage I and the causal explanations are realistic, except for one addition: it has become clear that the paper company will not be able to guarantee continuous high level activity in the forest because of fluctuations in the international market for paper.

Various design alternatives are generated by the company's technical people, based on different road plans and harvesting technology. The social analyst and her discussion group take it upon themselves to develop alternatives for the organisational side of the project. A number of new alternatives are developed, based on the idea of sharing the few full-time jobs between a large number of farmers and on persuading the company to operate on a lower level of activity, regardless of international market conditions. This lower output could be achieved with simpler equipement, but the hourly wage would have to be dropped for the scheme to be profitable.

The company finds the idea of job sharing acceptable, given the lower wage rate, and so do the farmers, because of the chance of avoiding social conflict in the small community. "After all", the community council man points out, "the income of the community as a whole will increase compared with the income level without the project, even though the harvesting volumes and wages rates are lower".

Finally, the appraisal finds itself with three road alternatives, two technological alternatives, three job-sharing plans, and a range of four possible harvesting levels. The discussion group has no opinion on which road access is best, but excludes one harvesting technology because it would require technical training that few of the farmers would be able to acquire. It also excludes one job-sharing plan alternative because it would prove too disruptive. These decisions make two of the harvesting levels unachievable, and so the group is left with twelve alternatives. They decide to leave the choice of road design to the technical people and to agree on the following ranking of the four remaining alternatives.

As wage rates can be adjusted, the economic assessment is still attractive. Environmentally speaking, the project is attractive so long as the fourth alternative is excluded, for it would unduly disturb a beautiful virgin stand in the outer valley. It is decided to proceed.

Table 3.2. DISCUSSION GROUP RANKING
OF VARIOUS PROJECT ALTERNATIVES

Alternative	Base period in work rotation scheme	Harvesting level (m^3/yr)	Ranking by discussion group
A	1 week	50 000	1
B	1 week	75 000	4
C	14 days	50 000	3
D	14 days	75 000	5
"Without"	6 months	40 000	2

Table 3.3. SOCIAL EFFECTS OF VARIOUS PROJECT ALTERNATIVES

Social effect and exposed group	Project alternative				
	A	B	C	D	"Without"
Reduced employment (Hits: old farmers)	2	1	2	1	0
Pressure to migrate (Hits: farmers without forestry jobs)	1	0	1	0	0
Monotonous working conditions (Hits: forestry workers)	1	2	2	2	0
Conflicts between local groups (Hits: whole community)	1	1	2	2	0
Instability in employment (Hits: whole community)	0	1	0	1	0
Disruptive lifestyle (Hits: forest workers)	2	2	1	1	0

0 = No effect.
1 = Some effect.
2 = Strong effect.

At the beginning of the Stage III appraisal of the few remaining alternatives, Table 3.3 of social effects is prepared to indicate who will be affected in what way. Social effects that are identical regardless of the chosen alternative are omitted.

On the basis of this table, and with the help of the graph in Figure 3.1 (established by the technical experts at the request of the discussion group), the social analyst writes a social-effects scenario for each alternative.

Starting with alternative A, with its low harvesting level and week-long job rotation scheme, she calls attention to reduced job opportunities (shaded in Figure 3.1) and the out-migration that could result. In her view, the near 50 per cent reduction in employment opportunity could lead to the forced exodus of at least one quarter of the farmers. The exodus would probably not occur till after construction and the first years of relatively inefficient use of the sophisticated equipment. The assumption that one half of the unemployed would move came from the discussion group. The group made this rough estimate on the basis of knowledge of the people involved, and on the fact that a community an hour away desperately needed forest workers and expected to do so even more in the future. This neighbouring community was willing to give people small landholdings to have them come and work in their forests. The social analyst therefore recommends polling all the farmers to get a better estimate.

She continues by describing the new work rotation scheme that requires the farmer to work for a period in the forest and then for a period at home. She cites some sociological

Figure 3.1. TIME PATH OF FORESTRY EMPLOYMENT
FOR VARIOUS PROJECT ALTERNATIVES

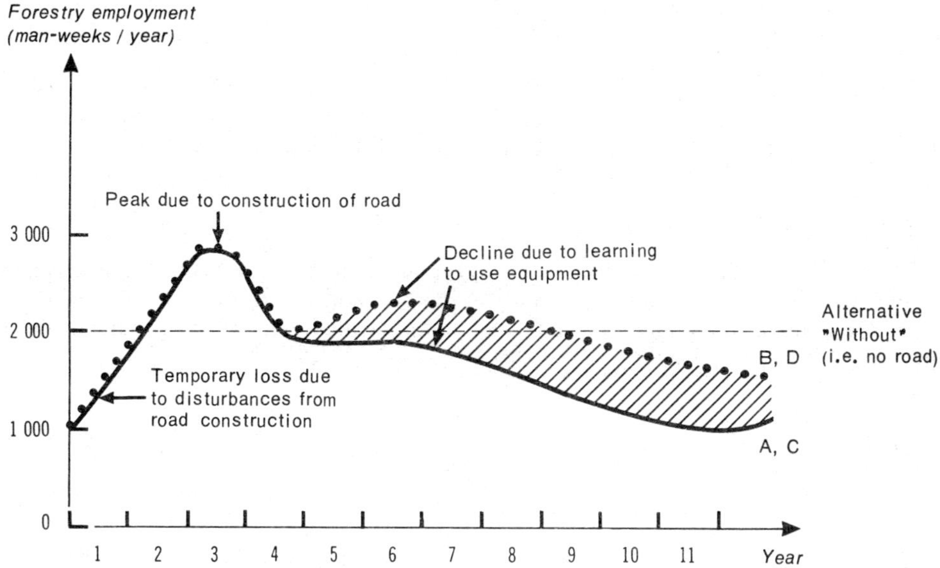

studies indicating positive experience with this type of variation in other job situations. But she reflects the opinion of some farmers by pointing to the expected difficulty in getting anyone to work in the forest during sowing and harvesting periods. Finally, she stressed the contribution to local stability that would result from the company guarantee to maintain removals at 50 000 m^3/yr, for at least ten years.

Once scenarios for the other four alternatives are completed, the material is presented in a 25-page pamphlet that is widely distributed in the area (through the discussion group members). After weeks of hectic public debate, the discussion group asked to rank the alternatives. As it turns out, the scenarios have not affected the opinion of the members at all.

However, the pamphlet has made the local community increasingly aware of the unemployment threat, particularly in the long run. Further there is an increasing tendency among the farmers to say that they would rather have a high and somewhat uncertain activity level than a low but safe harvesting rate. They feel that the high alternative at least holds out a chance of maintaining the traditional number of jobs, whereas the low alternative seems to guarantee out-migration within the next decade. Finally, it is decided to take a general vote, including all community members over 18 years of age.

The appraisal team draws up the results, shown in Table 3.4. This is attached to the scenario pamphlet along with the discussion-group ranking. Some words are added explaining the assumed rationale behind the vote, describing participation rates and, to the extent known, what groups voted for which alternative. These pages represent the final social appraisal of the

Table 3.4. COMMUNITY RANKING OF PROJECT ALTERNATIVES

Alternative	Number of vote	Rank (by voters)
A	290	3
B	350	1
C	70	5
D	90	4
"Without"	330	2

Inner Valley Road Project. It is combined with the economic and environmental assessments, and the final report is distributed to the project initiator and to the county council which will have to make the formal approval if the paper company finally decides to go ahead with the project.

NOTE

1. For example, it would be double counting to include "employment" as a beneficial social effect if one does not think that there are any benefits accruing from having a job other than that of receiving a paycheck. Most people believe there are such benefits: dignity, stability, a sense of participation.

Chapter IV

ENVIRONMENTAL EFFECTS

Two characteristics of forestry projects make appraisal of their environmental effects particularly important and often difficult. First, forests are inseparable from their environment: trees cannot be viewed in isolation from the soils they grow in, the watershed area they protect, or the economic and cultural life that has evolved around the forest. Consequently a broad perspective is needed if one wants to cover all effects of a forestry project. Second, forests develop slowly compared with most economic and technological processes. The best oak forests in Western Europe are managed over cycles which exceed 200 years; several centuries are required to reconstitute a degraded soil. Consequently, a very long term perspective is necessary to describe the full impact of a forestry project on the environment.

Needless to say, it is difficult, if not impossible, to achieve both a sufficiently broad and long-term perspective to reach a complete description of project effects. In practice, the appraisal team often has to restrict itself to the more manageable task of identifying "irreversibilities", that is, to try to find out whether a proposed alternative has effects that are not reversible during an acceptable interval – say 100 years – thereby reducing the options open to future generations in an irreversible manner.

If such irreversibilities are identified, they should be highlighted and fed back into the next round of project redesign. In many cases, the reasonable procedure is to drop an alternative with obvious negative irreversible effects, in favour of alternatives which maintain future flexibility, or, in other words, options which allow for adaptation to usages that are not foreseeable at present[1].

Many forestry projects are designed to specifically produce environmental benefits, such as erosion control, plant and animal protection, and so forth. Since the benefits are more obvious than the generally negative side effects of most forestry projects, we shall concentrate here on the less obvious negative effects. Needless to say, an appraisal team would present both negative and positive effects in their Report.

A. Questions Decision-makers Need Answers to

When trying to get a clearer picture of the environmental effects of a forestry project, the following questions are of particular importance:

1. What are the likely effects of the project on soil, water, air, vegetation and animals?

   In the description of environmental effects one should concentrate, though not exclusively, on effects that are not covered by the economic and social appraisals[2].

This usually requires the assistance of environmental experts since other people rarely have the ability to foresee ecological side effects and/or long-term effects of a project. Such assistance will help avoid the common excessive emphasis on short-term economic and employment effects.

2. What are the likely irreversible effects of the project?

   Needless to say, it takes great skill to give a good answer to this question. Still, it is possible to learn a great deal by looking at what has been done in past projects.

3. How uncertain are the answers to questions 1 and 2?

   Normally, there will be great uncertainty about predictions on environmental effects. Moreover, the uncertainty cannot be reduced by further study, since the effects may not occur for several decades. The reasonable procedure seems to be to identify and describe the uncertain effects and, in the next round of project redesign, to select those alternatives which present the least damaging or critical uncertainties.

B. Procedure During Each Stage of the Appraisal Process

As is the case of other effects, the assessment of environmental effects should progress by stages. Each stage covers the same activities but the amount of detail increases from one stage to the next.

At each stage, the procedure can be as follows:

- tabulate (for example, by using checklists) possible effects of the project on different components of the environment;
- draw up an "impact matrix" indicating the strength of the different effects for the various alternatives;
- describe the likely development over time of the different effects, trying to explicitly depict the likely situation in the distant future;
- use these results to establish, possibly in co-operation with the decision-maker, a classification of project alternatives according to their effects on the environment or, if possible, a ranking of the alternatives;
- propose means to reduce negative effects and strengthen positive effects through project redesign.

In contrast to social assessment, environmental assessment preferably should be done by specialists and not by a local interest group. The reason is that many environmental effects are obscure "systems effects"; laymen are not aware of the many interactions within the forest itself and between the forest and the environment. These relationships should be considered at any given moment, as well as over time.

Those with experience in environmental assessment best appreciate the importance of the "systems approach" and the notions of cycles and thresholds. There are two forms of cycles: first, closed streamflows of oxygen, water, carbon, nitrogen and various other mineral elements which pass through all the components of the ecosystem. Second, there are the biological cycles of growth, nutrition and reproduction of each plant or animal species. Both cycles must function for the ecosystem to remain unimpaired.

A threshold is a critical value of an ecological parameter. If the value of the parameter crosses the threshold, then a characteristic relationship of the ecosystem is destroyed and an important change, which is often irreversible, appears in the way the ecosystem functions. A threshold is not an absolute value determined independently of the other ecological parameters. On the contrary, it is a relative value that depends mainly on the characteristics of the ecosystem under consideration. Thresholds must therefore be examined case by case.

Stage I: Very Rough Environmental Appraisal of the Project Idea

This first, very rough appraisal of the project idea is designed to present the idea to an individual with ecological knowledge and insight, ideally from experience with similar projects. The objective is to get his immediate reaction and his comments about the likelihood of unexpected major and/or irreversible environmental effects. The first, very rough appraisal consequently consists of little more than a telephone call, or a short visit by the project initiator to a nearby ecologist.

Stage II: More Detailed Environmental Appraisal for Screening Purposes

Stage II should focus on a census of the main effects of the project, establishment of an impact matrix and a very rough indication of the long-term development.

Not too many effects should be chosen for presentation to the decision-maker, probably not more than five. The most significant are those that express effects which are particularly important for the environment and those that make it possible to distinguish between project alternatives.

The following list of environmental indicators is obviously not exhaustive.

- one or two "soil" indicators, such as erosion and fertility;
- one to three "water" indicators, like quantity, quality, likelihood of calamity;
- an "air" or "climate" indicator, if necessary;
- two or three "flora" and "fauna" indicators, e.g. productivity of the animal or vegetable biomass, diversity of the ecosystem, protected species, likelihood of fire damage;
- an indicator of the quality of the landscape.

In addition, the rough appraisal should point up possible risks of irreversibility in the ecosystem and discuss the capacity of the forest to adapt to changes under long-term economic or social conditions.

The Stage II appraisal ought to be conducted by ecology experts whose task is to eliminate project alternatives with grave environmental effects and to highlight alternatives with positive environmental effects.

Stage III: Thorough Environmental Appraisal of the Most Promising Project Alternatives

Here, much greater effort should be made to identify likely environmental effects, their strength and especially their likely development over time. The written scenario for each alternative can be illustrated, for example, with maps showing the extent of denuded areas at various points in time, with graphs showing the undiscounted expected development of, say, necessary clean-up costs to maintain water quality, or with drawings showing what it will look like at various times in the future.

In selected cases it could be worth mapping the complicated webs of cause-and-effect to help trace likely developments over time. And in even fewer cases, it may prove helpful to formalise the system's view in a computer simulation model to trace the consequences of various project alternatives.

Stage III appraisal obviously is a task for specialists with a good ability to distinguish between what is important and unimportant. Most of the existing Environmental Impact Statements are impenetrable collections containing so much data that the volume is useless as a practical guide to decision-making.

C. Examples of Environmental Effects

Having set out the procedural principles that make it possible to reply to decision-makers' questions, and having indicated how to develop them at the different stages in the appraisal process, we shall now furnish a few examples of environmental effects and discuss how they can be handled.

1. Effects on the Soil

The soil constitutes one of the basic elements in any forest ecosystem: on one hand, the value and quality of a forest directly depend on the fertility of the soil which supports it; on the other hand, soil development depends on plant cover and climate. The microclimate governing the soil is itself greatly modified by the presence of the forest. The soil encloses all the main biological cycles: the cycles of the main mineral elements (nitrogen, phosphorus, sulphur, etc.) are possible only because of the presence of the microflora and microfauna of the soil and particularly of decomposing organisms. The soil also plays a very important role in the water cycle. One of the essential roles of the forest is its capacity to conserve, enrich and even create soil.

Certain forestry projects have as their primary objective soil conservation and enrichment. In other types of projects, the problem of soil conservation appears not as a primary objective but as a constraint that must be respected by means of different silvicultural techniques. Some examples are the introduction (or maintenance) of a deciduous understorey in certain coniferous forests in order to improve humus quality; the banning of certain reafforestation species that might acidify the soil in an irreversible manner; the abandonment of certain techniques when they are likely to cause substantial erosion.

In order to anticipate effects on the soil, one must often rely solely on empirical rules based on experience, for scientific observation of the soil is rare and is considered relatively complex and costly. But when collected, soil data provide a precise description of the initial state of the soil (soil type, nature, texture, structure, composition of the various horizons of the soil profile, mineral content, humus content, etc.). It is, however, much more difficult to anticipate the changes in different parameters, e.g. migration of mineral elements, change in organic matter and humus, modification of the different horizons, etc. In some special cases, it can be worth the effort to monitor soil development at fairly long intervals (five-year spans, for instance) to provide empirical information on the adverse effect of certain silvicultural practices or the positive effect of others.

Large scale soil analyses are usually prohibitive, but in many projects a simplified soil analysis can help in the Phase II screening to eliminate alternatives which are incompatible with the existing soil. Whether very detailed or simplified, the data gathered is couched in specialised language that is often incomprehensible to the decision-maker, and should be "translated" to facilitate choice between alternatives.

For instance, erosion can be appraised in terms of a quantitative indicator: the number of tons of soil eroded per hectare per annum; or, more descriptive, the number of millimetres of soil removed. If, as is often the case, precise knowledge is not available, one is left with a qualitative indicator of the type: "erosion risk very high; high; medium; low or nil";

"reconstitution speed low; medium; high; very high". This scale may also be expressed by numbers, for example, from – 4 to + 4.

The various effects relating to humus and mineral content can be summarised in a "fertility" indicator. In the absence of a universal unit of measure for fertility, a qualitative scale of the following type can be used: "high fertility increase, medium increase, slight increase, unchanged, slight deterioration, medium deterioration, high deterioration".

When the water-retaining capacity of the soil is of importance, this effect could be included in the fertility indicator, or be assigned a special quantitative or qualitative indicator.

2. *Effects on Water*

Forests affect water systems in numerous ways: modification of the microclimate under the forest canopy and, in certain cases, of the mesoclimate; local changes in infiltration and run-off; changes in the water balance through evaporation and transpiration; changes at watershed level in streamflow and ground water recharge; indirect changes through the soil; transport of suspended solids due to erosion; presence of mineral salts due to leaching or of organic decomposition products.

Very elaborate water models exist to assess these various effects, but they are seldom available, or the necessary data is too costly to obtain. Often one must make do with approximate data based on the observation of similar projects. Table 4.1 summarises typical effects of various projects on water resource characteristics. Other possible descriptors of effects on water are:

- average quantity of water supplied by a given watershed expressed in m^3/year;
- water quality expressed by various quantitative indicators (temperature, oxygen content, content of various dissolved mineral salts, content of suspended solids, content of pathogens, etc.) or by a global qualitative indicator.
- flooding risk (expressed as decennial or centennial flood height at some points(s) of the watershed: town, industrial zone, etc.);
- low water risk that could cause economic damage;
- reservoir sedimentation rate.

In cases where the water effect is central, approximate data may not be acceptable, and it will be necessary to carry out large-scale measurements and construct very elaborate models. Ex post monitoring of the projects will then be of interest in order to provide new data and improve the models.

3. *Effects on Air and Climate*

The forest canopy has important effects on the microclimate and obvious effects on the evolution of the soil, water resources, and the composition of flora and fauna. On the other hand, its influence on the mesoclimate is more difficult to ascertain. Effects on the air may include the following: noise abatement; capture of various may include the following: noise abatement; capture of various pollutants; volatile emissions (some of which may be beneficial to health); and especially the conversion of carbon dioxide into oxygen. In the long term, of course, the forest is an overall oxygen producer only if the carbon fixed by chlorophyl photosynthesis in the organic matter is not reconverted into carbon dioxide by wood burning (forest fires or use of firewood) or by natural biological decomposition (as when an unexploited forest reaches a "climax" stage).

In many cases the effects on air and climate can be taken into account indirectly through the indicators for soil, water, flora and fauna. However, it can sometimes be desirable to

Table 4.1. RELATIVE EFFECTS OF WATERSHED ACTIVITIES ON WATER YIELD CHARACTERISTICS

Watershed management and related land use activities	Quantity of water yield	Soil loss by erosion	Water yield and related characteristics*					Flood risks	Frequency of low-water
			Water quality indicators				Disease org. and coliform		
			Temperature	Oxygen content	Nutr. elements	Sediment			
1. Grazing	+	+	0	–	+	+	+	+	+(?)
2. Timber cutting :									
Thinning	+	+	?	?	?	?	0	+	+(?)
Clearing	+	+	+	?	?	?	0	+	+
3. Road construction	+	+	?	?	+	+	0	+(?)	+
4. Herbicide applications	+	?	+(?)	?	+	?	0	?	+
5. Fire	+	+	+	–	?	+	0	+	+
6. Recreation use	?(+)	+	?	–	+	+	+	+(?)	+
7. Conversion from trees or brush to grass	+	+(?)	+(?)	?	?	?	0	+	+
8. Conversion from grass to brush or trees	–	–	–	?	?	?	0	–	–
9. Urbanisation	+	+	+	–	+	+	+	+	+

Likely effects : + = increase ; – = decrease ; ? = questionable effect, and "0" indicates no effect expected.
Source : H.M. Gregersen and K. Brooks, 1980 "Economic Analysis of Watershed Projects", FAO *Forestry Paper No. 17* – Supplement 2, Rome, FAO.

present an indicator of "air quality" or "climatic amenity". This could apply when forestry alternatives are to be compared with other kinds of land use (agriculture, industry, etc.).

4. *Effects on Flora and Fauna*

Through its effects on the microclimate and the soil, the canopy of the dominant storey of a forest stand exerts a decisive influence on the floristic composition of the other storeys. In turn, the composition of the flora influences the wildlife composition.

There are two aspects of the composition of flora and fauna that should be given special attention: conservation of the genetic heritage and the stability of the ecosystem.

Conservation of rare or threatened species is often an issue in forestry projects, but even for the most abundant species, it may well be desirable to conserve a diversified gene pool. Individuals living at the boundary of the area of their species possess in their genome special capacity of adaptation, and therefore are particularly important in maintaining the diversity of the genetic heritage of their species.

The more diversified an ecosystem, the greater its stability[3]. A forest will therefore better withstand the various ecological disequilibria, and notably attacks by animal or vegetable pests if it has a more diversified flora and fauna. Ecological diversity helps protect wood production against plant pathology problems.

Similarly, the danger of fire is connected with the composition of the flora: in the vast majority of cases, fire needs brushwood and low ligneous plants in order to develop and catch the tree-tops. When a forest canopy is dense enough to prevent the growth of brushwood or to maintain a sufficiently moist microclimate, then fire will normally only be an ordinary running fire which does not destroy the dominant storey of the forest.

For describing the likely effect on flora and fauna, a variety of methods are available, including the following:

- study of plant associations which makes it possible to predict the subsequent evolution of the canopy. The last stage of the "natural" evolution is called the "climax". Determination of the plant association present in a given place gives an indication of the "distance" from the climax and of the likely evolution of the vegetation;
- study of animal biocenoses, and of trophic chains;
- measurement of animal or plant biomass;
- study of the cicatrisation rate of the vegetation which indicates its sensititivity to different types of aggression;
- study of the specific ecology of certain animal or plant species.

The use of these methods (often reduced to simple diagnosis in the field) can form the basis for a number of indicators, for instance:

- primary production in tons of plant biomass produced annually per hectare. This will often be one of the most interesting and reliable indicators for appraising the very long-term economic value of the forestry project;
- ecosystem diversity, expressed according to a Shannon-type indicator or a qualitative scale;
- "distance from climax" deduced from the study of the vegetation sequence and expressed along a qualitative scale of the type "climax stage, near climax, far from climax, very far from climax";
- number of rare or endangered species present on the site;
- a global indicator of "ecosystem quality" along a qualitative scale can summarise and replace the above indicators in a more "eloquent" fashion;

- one or more indicators of forest sensitivity to various forms of aggression: fire; climatic accidents (drought, frost); parasitic attack (insects, fungi). These may be aggregated in a global indicator of forest sensitivity (or conversely forest stability).

5. Effects on Landscape Value

The composition of the flora and fauna determines the quality of the forest landscape, the quantity and quality of game and fish, through water quality. In this context, the landscape value not only means the visual setting of the forest, but also includes all sensations perceived by the forest walker: sounds, smells, sensations of warmth, touch, etc.

The effect of a forestry project on landscape value stems from: the recreation potential of the forest, which depends on microclimate (temperature, moisture, wind) and composition of flora and fauna (absence of offensive plants and animals, undergrowth hampering walking, etc.); and the setting provided for sites of historic or artistic interest.

It may be useful to have landscape planners study, with their specific techniques, the effects of the project on the landscape in order to arrive at a "landscape quality" indicator for presentation to the decision-maker. In some cases, it may be desirable to provide models or sketches illustrating these effects. This part of the appraisal often overlaps directly with the social and economic assessments.

NOTES AND REFERENCES

1. de Montgolfier, Jean and Bertier, Patrice. "Tentative pour poser le problème en termes de patrimoine", *Revue forestière française*, numéro spécial 1980, pages 115-125.
2. See comment in Chapter II concerning double counting.
3. Ashby's Law of Requisite Variety.

Chapter V

ECONOMIC AND FINANCIAL EFFECTS

Projects have economic and financial effects that can be measured in terms of monetary values, based either on market prices or shadow prices[1]. All the principles of social and environmental impact assessments apply here as well. Thus, the "with and without" principle, i.e. consideration of only the incremental values associated with the project, applies directly to economic and financial assessments. Similarly, consideration of separability of project components and the appropriate forms of analysis for given project situations and components within a project are of great importance. Lastly, proper treatment of uncertainty is essential in looking at financial and economic impacts.

A. Questions Decision-makers Need Answers to

In most cases, the following five questions are the main economic and financial ones considered by decision-makers in project appraisals:

i) Is the value of the benefits greater than the value of the costs of the project, when both are measured as economic values (in monetary terms) and appropriately adjusted to take time into account? This is the basic question of "economic efficiency". When one of the objectives sought is to increase the aggregate economic benefits (goods and services available) derived from the use of the nation's limited resources, then this is a relevant question.

ii) What are the distributional impacts of the projects, i.e. how are the economic benefits and costs of the project distributed among the different groups concerned? This boils down to the question of who pays and who gains? OECD Member countries are concerned with just such income redistribution effects of public investments, i.e. the equity issue. In most projects, concern centers on different income classes (do the poorer members of society receive more benefits than the richer members?), and regional groups (does region X receive more benefits than region Y?).

iii) How will the project affect economic stability? This question relates generally to the foreign exchange impact of the project in relation to balance of payments. In countries with a large imbalance, this question is of critical importance. Other issues of importance here are regional income and job stability. For example, a recreation project of a highly seasonal nature may involve a great deal of instability in terms of income flows; or a timber harvesting project that does not adequately take into account sustainable timber production over time may have an initial positive impact on incomes and economic activity, but eventually result in hardship and instability as the level of economic activity developed intially can no longer be supported by available timber.

Finally, several very pragmatic questions of interest to decision-makers can be mentioned under the heading of economic and financial impacts:

iv) Does the project stay within the budget limitations of the government agencies involved? The traditional budget analysis that is incorporated into most project appraisals attempts to answer this question. Obviously, if budget limits are exceeded, then the project will have to be readjusted to be feasible.

v) Is the project financially attractive to all the private entities that may volunteer funds for it? Private entities can be involved in such projects either through cost-sharing or concessions, and if they do not find the project attractive enough, then it may be necessary to consider obtaining subsidies.

Procedures have been perfected for treating each of these five questions, from the initial rough appraisal to the final detailed feasibility studies. As is the case of the other project effects, it is important to remember that economic and financial assessments should be explicitly incorporated in the appraisal process from beginning to end so as to avoid unnecessary expenditures of funds on analysis of project alternatives that for one reason or another can in fact be eliminated early on.

For example, let us assume that an idea exists for a major forest recreation project. A rough calculation of the gross budget requirements shows that the project as initially proposed exceeds the budget allocation of the agency considering the project. This indicates that the project needs to be redesigned before a lot of time and funds are spent analysing detailed social, environmental and economic impacts of the project.

B. Procedure for Answering the Questions

There is a well defined procedure for each of the above questions, regardless of the stage in the appraisal process. The differences in how this procedure is carried out at different stages relate to the sophistication of the estimating techniques used, the amounts of data required and collected, and the level of detail considered and time devoted to the analysis.

The steps involved in answering the five questions are shown in summary form in Figure 5.1. Usually, they are simultaneously to gradually develop a set of useful tables called "interrelated working tables".

To illustrate the nature of such tables, we will use a simple, hypothetical example of a project that involves planting some 2 000 ha of marginal farm land owned by about 200 private individuals. The government is entering the project by providing technical assistance, cash subsidies for planting, and the planting stock. From the government's point of view, the total planting programme is duly treated as a "project".

Given the above, the following set of interrelated tables could be prepared (details on how entries in the tables were derived would be shown in appendices in the appraisal reports):

1. A table showing physical inputs to and outputs from the project over time (e.g. inputs of labour, land, materials, machinery and outputs of timber, recreation days, etc.). See Table 5.1 for an example.
2. A set of tables showing disaggregations of inputs and outputs in terms of relevant groups that will supply inputs and/or receive benefits from outputs. Samples are given in Tables 5.2 - 5.5.
3. A table showing expected market prices for those inputs and outputs which are or will be bought and sold in markets. Prices are given in real terms (i.e. adjusted for inflation if different points of time are involved; an example is given in Table 5.6).

Figure 5.1. MEASURES OF PROJECT WORTH:
NET PRESENT WORTH AND INTERNAL RATE OF RETURN

1. *Net present worth**

NPW is a measure of the present (discounted) value of the net benefit flow of a project. It is the sum of the present value of all benefits minus the sum of the present value of all costs *or* the sum of present values of benefits minus costs for each period *t*.

More formally, it can be written as follows:

$$NPW = \sum_{t=0}^{n} \frac{B_t}{(1+i)^t} - \sum_{t=0}^{n} \frac{C_t}{(1+i)^t}$$

or

$$NPW = \sum_{t=0}^{n} \frac{B_t - C_t}{(1+i)^t}$$

where :

B^t = benefits in each year t
C^t = costs in each year t
n = number of years to end of project
i = discount rate

2. *Internal rate of return (IRR)*

The internal rate of return *(IRR)* can be defined as:

- the rate of discount which makes the present value of costs equal the present value of the benefits, or the $NPW = 0$;
- the average earning power of invested funds or the average rate of return on the invested funds outstanding per period while they are invested in the project.

IRR is that rate of discount such that:

$$\sum_{t=0}^{n} \frac{B_t}{(1+IRR)^t} = \sum_{t=0}^{n} \frac{C_t}{(1+IRR)^t} = 0$$

or

$$\sum_{t=0}^{n} \frac{B_t - C_t}{(1+IRR)^t} = 0$$

where terms are as defined earlier.

* Also referred to as "present net worth" or "net present value".

4. A table showing shadow prices for all those inputs and outputs for which shadow prices have to be developed. The selection will generally depend on practices in force for different agencies/governments. (See Table 5.7).
5. A total "value flow" table for the project, developed by combining information from Tables 5.1 and 5.7, as shown in Table 5.8.
6. A set of cash flow or value flow tables (whichever is appropriate) corresponding to Tables 5.2 through 5.5. See Tables 5.9 through 5.12 for examples. Where appropriate, taxes, subsidies, and loan entries have been included.

Table 5.1. PROJECT PHYSICAL INPUTS AND OUTPUTS

Item	Units[a]	Years 0	1	2	3-14[b]	15
Inputs						
Unskilled labour	1 000 m.-d./year	4	2	2	1	1
Semi-skilled labour	1 000 m.-d./year	4	1	1	1	1
Skilled labour	1 000 m.-d./year	2	1	1	–	–
Supervision	m.-d./year	40	30	10	10	10
Extension	m.-d./year	400	100	100	40	200
Tractors	Machine days/year	2 000	30	30	10	10
Trucks	Vehicle days/year	440	130	110	50	210
Land[c]	Ha.	2 000	2 000	2 000	2 000	2 000
Seedlings	1/000 seedlings/year	2 000	–	–	–	–
Outputs						
Pulpwood	1 000 m^3/year					320
Firewood	1 000 m^3/year				10	

a) m.-d. = man days.
b) Items in this column are *per year* during period shown.
c) The same land is of course used every year.

Table 5.2. FOREST AGENCY–INPUTS AND OUTPUTS

Items	Units[a]	Years 0	1	2	3-14[b]	15
Inputs						
Supervision	m.-d./year	40	30	10	10	10
Extension help	m.-d./year	400	100	100	40	200
Seedlings[c]	1/000 seedlings/year	2 000	–	–	–	–
Vehicles	Vehicles days/year	440	130	110	50	210
Outputs		None to public sector				

a) m.-d. = man-days.
b) Items in this column are *per year* during period shown.
c) Seedlings are given to landowners by the Agency.

Table 5.3. PRIVATE LANDOWNERS – INPUTS AND OUTPUTS

Items	Units	Years 0	1	2	3-14[a]	15
Inputs						
Labour[b]	1 000 m.-d./year	10	4	4	2	2
Tractors	Machine-days/year	2 000	30	30	10	10
Lands	Ha	2 000	2 000	2 000	2 000	2 000
Outputs						
Pulpwood	1 000 m^3/year					320
Firewood	1 000 m^3/year				10	

a) Items in this column are *per year* during the period shown.
b) Breakdown might be made similar to that in Table 5.1.

Table 5.4. ADDITIONAL DIRECT EMPLOYMENT BY INCOME GROUPS
Man days/year[a]

Income groups	Years				
	0	1	2	3-14	15
\$0 – 4 000	4 000	2 000	2 000	1 000	1 000
\$4 000 – 6 000	4 000	1 000	1 000	1 000	1 000
\$6 000 – 15 000	2 000	1 000	1 000	–	–
\$15 000 +	40	30	10	10	10

a) Similar tables could be prepared showing indirect employment effects.

Table 5.5. IMPORTED INPUTS AND EXPORTED OUTPUTS AT DIFFERENT TIMES[a]

	Units	Time (year)	Volume
Imported inputs			
Trucks	Trucks/year	0	3
Tractors	Tractors/year	0	3
Exports			
Pulpwood	1 000 m^3/year	15	320

a) Import substitutes would also be included in this table.

Table 5.6. ESTIMATED MARKET PRICES[a]

Item	Units	Price
Inputs		
Unskilled labour	\$/m.d.	20
Semi-skilled labour	\$/m.d.	30
Skilled labour	\$/m.d.	40
Professional	\$/m.d.	60
Tractors	\$/machine day	200[b]
Trucks	\$/vehicle day	40[b]
Land	\$/ha. year	10[c]
Seedlings	\$/1 000 seedlings	50
Outputs		
Firewood	\$/m^3	10
Pulpwood	\$/m^3	20[d]

a) In real terms, i.e. in fixed dollars per unit (e.g. 1980 dollars) if different times are involved.
b) Including operating cost. Price of equipment based on import value.
c) Entered as rental equivalent.
d) Based on FOB price, adjusted to take into account transport and handling costs from forest.

Table 5.7. SHADOW PRICES[a]

Item	Units	Shadow price
Inputs		
Unskilled labour	$/m.d.	15
Semi-skilled labour	$/m.d.	30
Skilled labour	$/m.d.	40
Professional	$/m.d.	60
Tractors	$/machine day	200
Trucks	$/vehicle day	40
Land	$/ha. year	10
Seedlings	$/1 000 seedlings	50
Outputs		
Firewood	$/m³	10
Pulpwood	$/m³	20

a) Since only unskilled labour has a shadow price different from market traded price in our example, and since no non-market traded inputs and outputs were identified, Table 5.7 could have been eliminated. We have included it here for the sake of clarity.

Table 5.8. TOTAL ECONOMIC VALUE FLOW TABLE

$ 1 000/year[a]

Item	Years				
	0	1	2	3-14	15
Costs					
Unskilled labour	60	30	30	15	15
Semi-skilled labour	120	30	30	30	30
Skilled labour	80	40	40	–	–
Supervision	2.4	1.8	0.6	0.6	0.6
Extension	24.0	6.0	6.0	2.4	12.0
Tractors	400	6.0	6.0	2.0	2.0
Trucks	17.6	5.2	4.4	2.0	8.4
Land	20.0	20.0	20.0	20.0	20.0
Seedlings	100	–	–	–	–
Total	824	139	137	72	88
Benefits					
Pulpwood					3 200.0
Firewood				100.0	
Total				100.0	3 200.0
Net benefits (costs)	(824)	(139)	(137)	28	3 112

a) In real terms, e.g. fixed 1980 dollars.
Source: Tables 5.1 and 5.7.

Table 5.9. FOREST AGENCY OUTLAYS[a]

$ 1 000/year

Item	Years				
	0	1	2	3-14	15
Costs					
Supervision	2.4	1.8	0.6	0.6	0.6
Extension	24	6.0	6.0	2.4	12
Seedlings	100	–	–	–	–
Vehicles	17.6	5.2	4.4	2.0	8.4
Total cost	144	13	11	5	21

a) This table could also include *increased* tax returns to government due to the project. In that case, it would detail total government outlays/inflows, and not merely forest agency consideration.

Source: Tables 5.2 et 5.6.

Table 5.10. CASH FLOW TABLE FOR PRIVATE LANDOWNERS

$ 1 000/year

	Years				
	0	1	2	3-14	15
Revenues (receipts)					
1. Sale of products				100	3 200
2. Subsidy[a]					
Sub-total				100	3 200
Cost (expenditures)					
3. Labour	280	110	110	50	50
4. Tractors	400	6	6	2	2
5. Land[b]	20	20	20	20	20
Sub-total	700	136	136	72	72
Net revenue (cost)	(700)	(136)	(136)	28	3 128

a) Space for entry of subsidy figures, if one wants to study the consequences.

b) The land cost is entered as the rental equivalent. This is a cost in the sense that the landowner foregoes it by planting his land instead of renting it out or producing other crops on it.

Source: Tables 5.3 et 5.6.

Table 5.11. ADDITIONAL INCOME BY INCOME GROUPS

$ 1 000/year

Income groups	Years				
	0	1	2	3-14	15
0 – 4 000	80	40	40	20	20
4 000 – 6 000	120	30	30	30	30
6 000 – 15 000	80	40	40	–	–
15 000 +	26.4	7.8	6.6	3.0	12.6

Note: Supplementary tables could be used to show the average rate changes due to project..

Source: Tables 5.4 et 5.6.

The procedure for establishing these tables is the same, regardless of the stage in the project planning and appraisal process, except for the amount of effort put into the task. Next, it is useful to summarise how the tables are used to answer the five questions.

Question 1: Is the economic value of the benefits greater than the economic value of the costs of the project?

Two measures are commonly calculated to assist in answering this question: the net present worth and the internal rate of return. In comparing mutually exclusive alternatives, the appropriate measure of comparison is the Net Present Worth (NPW). For ranking projects in a situation where a limited budget is to be allocated among not-mutually exclusive alternatives (i.e. capital budgeting), an appropriate ranking measure is the Economic internal Rate of Return (ERR). Appendix I describes both of these measures and compares them in more technical terms. Since in most cases decision-makers are concerned both with the best design for a given project (mutually exclusive alternatives to achieve a given project objective) and with the problem of allocation of a limited budget to a number of projects, both measures should generally be calculated.

Table 5.8, derived from information in Tables 5.1 and 5.7, provides the necessary information for calculating both the NPW and the ERR. The calculation techniques are well-known and straightforward[2].

If the present value of benefits exceeds the present value of costs, then it will have a positive NPW, when the appropriate social discount rate is used to calculate present values, and it also will have an ERR that exceeds the social discount rate. Thus, either measure can be used to answer Question 1. However, if the objective is to screen alternatives that are mutually exclusive, then the net present worth measure should be used.

Question 2: What are the distributional impacts of the project?

There may be several dimensions to this question, depending on the project situation and the particular agency/government involved. Most often the question relates to how costs and benefits associated with a given project alternative affect different income groups, i.e. do poorer persons gain more than richer persons.

Table 5.11, derived from Tables 5.4 and 5.6, provides the necessary information to answer the question. In some cases, the table itself is all that is needed. In other cases, the information could be presented in comparative form, in a table or graph showing relative impacts of different project design.

There is no single acceptable measure of the effect of the project on income redistribution. In other words, it is not possible to put forward only one criterion for income distribution impacts that can be used to say whether the project is "good" or "bad" in terms of this question[3]. Rather, the decision-maker must be presented with estimates of impacts on different income groups over the life of the project.

Question 3: How will the project affect the balance of payments and economic stability?

As mentioned earlier, the main concern here is often related to the effect of the project on foreign exchange or balance of payments. In this case, Table 5.12, derived from Tables 5.5 and 5.6, provides the relevant information to answer the question. The table provides an indication of the net foreign exchange impact of the project alternative over time. What is judged to be acceptable depends on the particular circumstances. A number of single measures of foreign exchange impact have been developed to provide further guidance.

Table 5.12. BALANCE OF PAYMENTS EFFECTS
$ 1 000/year

Item	Time (year)	Foreign exchange	
		Outflow	Inflow
Tractors	0	120	
Trucks	0	45	
Pulpwood	15		3 200

Note: The inflow and outflow values could be discounted back to the present, using the social discount rate. This would provide a measure of the present value of direct net foreign exchange effect.
Source: Tables 5.5 and 5.6.

The basic question relates to whether or not the project uses more foreign exchange than it brings into the country. The main problem in answering the question is whether or not secondary foreign exchange effects should be included, i.e. the foreign exchange effects which are caused by the project, but not directly accounted for by imports of project inputs, by exports of project outputs, or by substitution of output for imports. For example, a given project may have a positive effect on the balance of payments in direct terms, but may result in large increases in imports in other sectors.

Assume a plantation project that will produce wood for an intended expansion of domestic pulp and paper production and some wood for export. There are no imports of inputs and some foreign exchange earnings through wood exports. Looked at in isolation, the plantation project has a positive impact on foreign exchange earnings. But assume that the pulp mill expansion involves large imports of foreign equipment and expertise. In this case, the secondary effects of the overall activity (plantation and mill) may have a negative effect on balance of payments. The question is: how far do we go in estimating the total effects of the project? The answer is that it depends on the situation in the country in question and the extent to which balance of payments is a critical consideration for decision-makers.

There is no single measure that can be recommended for measuring balance of payments impacts. The simplest approach is to provide the decision-maker with a table showing the direct net inflows (outflows) of foreign exchange associated with the project over time (see Table 5.12). A further step may be to discount the flows over time to provide a net present value of the effect. These types of tables include only the primary effects. Additional tables can be prepared showing estimated secondary balance of payments effects, if appropriate multipliers[4] are available. Such multipliers can be crudely estimated from previous experience, or they can be derived from input-output tables, if such are available for the country and/or region in question.

McGaughey (1979) suggests two additional measures which can be used to provide an approximation of the foreign exchange impact of a project. The first is a ratio of the net primary (direct) foreign exchange earnings per capital outlays for the project. The difference between the two is that, in the latter, the multiplier effects of the project outlays on imports and exports are included. The second thus provides a more complete indicator of project impacts on foreign exchange earnings.

Other stability impacts may also be of interest, for instance, effects on the pattern of jobs and employment. This is the main area where economic and social analyses overlap. It is an area in which there should be co-ordination of the project appraisal process to avoid duplication, but also to insure that all relevant impacts are covered.

Question 4: What is the budget impact and the net fiscal impact of the project for the agencies involved?

Table 5.9, derived from Tables 5.2 and 5.6, provides the necessary information to answer this question. Basically, this table provides information on the net inflow (outflow) of public funds, by year for the project. Simple accounting of the flows of public funds and the receipts by the government provide the best kind of information that can be presented to the public decision-maker. A financial rate of return (FRR) for the public sector (or the responsible agencies) could be calculated – but, in most cases, financial rates of return provide little guidance concerning desirability of public projects.

There is no other single measure that can be calculated to answer this fourth question. If an agency has budget guidelines set at the project level, then the information provided by a table showing the required financial contribution for the agency can be used to see if the project can be realised within existing budgets.

Decision-makers have to decide whether or not the proposed budget for the project is acceptable and how the project compares with other possible uses of agency funds. If the proposed project budget is too large, then the project can be redesigned on a smaller scale, or if that is not possible, it can be rejected as being not feasible under existing budget constraints.

Question 5: Is the project financially attractive to all private entities who voluntarily will have to put funds into the project?

Private companies and individuals are frequently involved in forestry investment projects. Table 5.13 provides examples of projects involving both public and private participation.

Table 5.10, based on Tables 5.3 and 5.6, together with an estimate of relevant alternative rates of return for the private participants involved, provides the necessary information for answering this question. (Alternative rate of return is the rate of return which private parties could get in the best alternative use of their funds). It may be that several tables will have to be developed for the different private entities involved.

Ultimately, the only way decision-makers will know whether or not the private participants are satisfied with the current project design is to ask them if they will participate

Table 5.13. EXAMPLES OF FORESTRY PROJECTS INVOLVING BOTH PUBLIC AND PRIVATE PARTICIPATION

1. Any project involving public *subsidies* to private entities (e.g. reforestation, conservation, industry).
2. Projects involving *sale of public timber or other forest outputs* to private users.
3. Regional forestry development or conservation projects *involving both public and private lands.*
4. *Concessions or leases* to private entities for use of public forest lands (e.g. ski resort or other recreational developments).
5. Public projects involving *contracts to private entities for work to be done* (e.g. road building, construction of facilities, tree planting).

and then to see how they actually react once the project is undertaken. However, at the appraisal stage, some rough indicators of financial attractiveness can be derived, if decision-makers have a general idea of the rates of return which are available to the participating private entities in alternative investments. This information can often be obtained through interviews and questionnaires, coupled with general information on alternative investments available to private entities.

Financial rates of return (FRRs) should be calculated for each entity that will have to make voluntary decisions on project participation. It is basically a simple process once cash flow tables are available.

The assumption is that if the FRRs are high enough, then private investors will participate, ensuring project feasibility for this particular area of the project.

If calculated FRRs for certain groups are below assumed alternative rates of return, then this information could provide a basis for upward adjustment of subsidies or public financial input into the project or modifications in laws and regulations. If higher subsidies cannot be justified, then a lower than acceptable FRR will indicate that risk capital is likely to be invested on the project. The project will either have to be redesigned or rejected in terms of this dimension of feasibility. If calculated FRRs for some groups are considerably above maximum needed rates of return, then this may indicate the possibility of downward adjustment in subsidy levels or increases in taxes, depending upon what other objectives (e.g. related to income redistribution) are relevant in terms of the group(s) considered.

C. Procedure at each Stage of the Appraisal Process

Now we can look at how the procedure outlined in Figure 5.1 and resulting in Tables 5.1 to 5.7 fits into the three stages of the appraisal process. Table 5.14 shows in summary form how the derivation of Tables 5.1 to 5.7 can be treated[5]. The following example shows how the process works in more detail.

A Sample Case: Afforestation of Marginal Farm Land

An initial project idea has been put forward to organise a forestry planting programme for some 2 000 hectares of abandoned farm land in Mythopia County. The forestry agency will have to take the lead in the programme, since private farmers do not have the necessary knowledge and experience. Furthermore, initial inquiries indicate that many of the farmers are not motivated to get involved in afforestation efforts on their own. Some feel that the profit potential is too low, while others simply lack the necessary capital. Afforestation would bring small profits in three years time, but the main results would be fifteen years away.

The project would involve forestry extension activities to acquaint the farmers with the project potentials and the technical requirements. It would involve cost-share subsidies from the government in the form of free seedlings and free technical assistance. Farmers would provide the labour and tractors needed for the planting and tending operations.

The reason for government involvement is to stimulate employment, provide productive use for presently marginal, abandoned farm land, and to provide cover on the land for watershed protection purposes. Moreover, the project would provide fuelwood for local communities, thereby relieving pressures on other types of fuel and on irrational exploitation of the existing forest for fuel. The pulpwood output from the project would be exported to a neighbouring country, where a ready market exists.

In the following discussion of the three stages in the appraisal process, only the economic and financial considerations are treated. The social and environmental impact analyses would have to be carried out in addition to the economic and financial assessments.

Table 5.14. PROCEDURE AT EACH STAGE OF THE APPRAISAL PROCESS

Step	Stage in process		
	I Rough appraisal of project idea	II More detailed appraisal for screening purposes	III Thorough appraisal of most promising project alternatives
Physical flows of inputs and outputs *(table 5.1.)*	Rough estimates of main categories of inputs and outputs listed, frequently based on rules of thumb.	More detailed breakdowns of inputs and outputs used. Some project specific estimates substituted for rules of thumb values.	Detailed breakdowns of required inputs and expected outputs, based on engineering and market studies.
Disaggregations of physical flows by those who supply inputs and receive outputs *(tables 5.2. – 5.5.)*	Rough tables showing main flows of inputs and outputs by groups of relevance. May be done in percentage terms at this point.	More exact disaggregations developped, based on more detailed specification of relevant groups, including traded items.	Original identification of groups that actually will be impacted by project. Market studies used to identify expected beneficiary groups.
Expected market prices *(table 5.6.)*	Market prices for main inputs and outputs ascertained. If price change projections (in real terms) available, these can be used to establish future prices.	More effort devoted to projecting market prices in real terms. Some additional distinctions made between inputs and outputs of concern.	Based on market studies, income, population, and price projections, detailed estimates of expected market prices (in real terms) are projected. Project effects on prices are estimated if possible.
Shadow prices *(table 5.7.)*	Generally, only labour and land shadow prices considered here on the input side; on the output side shadow price used only if acceptable ones exist.	Additional efforts devoted to determining appropriate shadow prices, particularly for non-traded, location specific impacts and inputs and outputs.	More detailed investigations of appropriate bases for shadow prices are carried out if deemed desirable. If nationally determined values are used, then additional work not needed.

Stage I: Very Rough Appraisal of the Project Idea

The idea for the project was initially suggested to the forestry agency by a local county forestry extension agent. He provided some very rough information on the potential plantable area, the farmers who would be involved, and the types of tree species which would grow best in the area for the given output objectives.

The first task was to put together an initial estimate of the physical input and output requirements for the project. These initial figures were developed on the basis of past experience and rules of thumb concerning inputs required to prepare and plant the type of land involved. At this stage, average figures for the entire 2 000 ha were used and Table 5.1 was prepared. This initial appraisal task was carried out by an experienced forester in the agency.

In addition to Table 5.1, the forester also estimated the contributions which would come from the agency and those which the farmers would provide. These estimates are shown in Tables 5.2 and 5.3. Since one of the objectives of the project was to provide employment and income for lower income groups, the appraiser also estimated the employment effects by income groups (see Table 5.4). At this stage, the estimates were only "guestimates" based on discussions with the extension forester who proposed the project and with local official in Mythopia county. Since another factor in the project involved exports of pulpwood, the appraiser provided some initial estimates of the exported outputs and required imports for the project (see Table 5.5).

The next step was to link the physical input and output information to dollar values, or prices. Based on past experience and local estimates of going wage rates and other cost elements plus wood-price trends, the appraiser prepared Table 5.6, which gives market prices for inputs and estimated prices for the output.

Recognising that, from the government's point of view, the project had a social objective, the appraiser quickly looked over the cost and return items to see which of them should be shadow priced. In this initial stage, only unskilled labour was deemed to need shadow pricing, recognising that the minimum wage requirements of the government overestimated the real social cost of employing the unemployed unskilled workers who would participate in the project. The minimum required market wage was $20 per man/day. The appraiser estimated that about three quarters of the unskilled persons who would be employed in the project were currently out of work and that the value of their production for home consumption which would have to be given up would be about $1.00 per man/day. Thus, a quick calculation – $\frac{1}{4} \times \$20 + \frac{3}{4} \times \$1$ – yielded the estimated shadow price for unskilled labour of $4.50 per man/day. All other inputs and outputs were considered to have shadow prices equal to their market values as shown in Table 5.7.

With the input and output information and value estimates in hand, the appraiser prepared Table 5.8, the value flow table for the entire project. Using the bottom line of this table and the discount rate of 4 per cent, provided by the agency as the relevant one sued by government, the appraiser calculated a net present worth (NPW) of $887.000 and an internal rate of return (ERR) of 8.7 per cent. Thus, at this initial stage, the project represented an economically attractive use of public funds, given the designated discount rate of 4 per cent. (The appraiser had been instructed that if the project did not meet the 4 per cent cutoff rate, she should stop work on the project).

Given this, the appraiser continued with the first stage appraisal. Since the agency's budget was limited, she developed Table 5.9, which provided information on the outflow of public funds needed to support the project over its life. She checked the resulting figures with the administrator and was told that the magnitudes were acceptable.

The next question was the desirability of the project from the point of view of the farmers who where expected to participate. Based on discussions with Mythopia bankers and agricultural extension personnel, it was estimated that the typical farmer could expect to earn on average 10 per cent per year (in real terms) on his capital investments (either in the bank or in agricultural investments).

With this information in hand, the appraiser prepared Table 5.10, which represents the financial cash flow for the project, using the average values discussed earlier. Using the bottom line of this table, a financial internal rate of return of 10.9 per cent and a net present worth of $95 000 were calculated for the project.

In other words, from the average farmer's point of view, the project would be seen as marginally profitable, given the estimated 10 per cent alternative rate of return. The appraiser noted this point in the Stage I report as an area that would require further investigation. The range of alternative rates of return available to farmers would have to be analysed, since if they turned out to be above 10.9 per cent, then farmers would most likely not be interested in the project. If that turned out to be the case, the government would have to consider the possibility of providing an outright cost-share subsidy to cover part of the site preparation and planting costs, thereby raising the rate of return from the farmers' point of view but also increasing the cost to the government.

Lastly, the appraiser made some estimates of the income effects by income groups (Table 5.11) and prepared a table showing the balance of payments effect of the project (Table 5.12).

Once this rough Stage I economic appraisal was completed, the project idea appeared to be sound. However, the social appraisal indicated that the income redistribution effects of the project might not be entirely satisfactory, since it would be primarily the larger and richer farmers who would benefit from the subsidies in the form of free seedlings and technical advice[6]. The environmental appraisal did not turn up any negative effects, and found several positive effects in terms of watershed protection and wildlife habitat.

Stage II: *Appraisal for Screening Purposes*

With these conclusions in mind, the forestry agency decided to continue with the project idea and to develop a Stage II appraisal that would make some adjustments for the income redistribution effect. For the Stage II appraisal, the part-time services of a forestry agency economist were secured, together with the services of two consultants, one a social scientist trained in social impact assessment. In addition, a soils expert was contacted for site productivity information, and several community leaders were informed of the project proposal and asked to serve as a contact with the local community.

An initial task was to disaggregate the farm population that was expected to participate into various groups to further clarify the project effects. A breakdown was made on the basis of farm size (which correlated with annual incomes). Within each of these groups, a further distinction was made on the basis of site classes for the potential project lands. Areas with less than three contiguous hectares of plantable land were eliminated from consideration on efficiency grounds.

Inputs and outputs were again disaggregated by public and private, but further disaggregation took place by ten income/site class groups. The total area available for these ten groups was 2 200 hectares. (Further investigation had indicated that the initial area estimate of the county extension agent was conservative.)

Availability of seedlings in sufficient quantity had been flagged in the Stage I appraisal as a major uncertainty. In Stage II, a survey of nursery stock in the project region indicated that the required seedlings would be available, but only over a three-year period. In order to meet this constraint, and given the forestry agency's somewhat cautious attitude, it was decided to phase the project planting activities over a three-year period, with 800 hectares the first year, 800 the second year, and 600 the third year. The assumption was that the first year's results would provide useful information for planning activities during the second and third years.

Further Stage II studies involved a more detailed investigation of labour availability during the required periods for site preparation, planting and tending. A preliminary market study for fuelwood was conducted to reduce uncertainty concerning markets and price expectations for this component.

Once the project appraisal team agreed on the relevant dimensions of the project and the timing and magnitudes of inputs and outputs by farmer groups, each of the team specialists started to develop Stage II information for the appraisal of particular effects. In what follows, we will trace only the activities of the economist.

Price assumptions for inputs and outputs were looked at again, particularly in terms of any expected relative increase over the project period. A particular concern was the effect that the project might have on prices, since it would represent a fairly substantial increase in fuelwood output in the region and since labour requirements were fairly substantial during certain years and periods during the year. Based on these analyses, several of the price assumptions (in Table 5.6 from Stage I) were changed to reflect expected effects of the project over time.

Similarly, the economist reviewed the shadow price assumptions of the Stage I appraisal. In this case, she found no need to develop additional shadow prices that differed from market prices for the direct inputs and outputs. However, based on rough estimates of erosion control effects associated with afforestation of some of the more steep, erodable lands within the project area, she prepared a brief analysis of the indirect benefits associated with the project, using shadow prices developed by the National Watershed Institute. These results were later combined with the direct wood-related benefits in the value flow table. Since there was no way of quantifying the dollar values associated with the identified wildlife benefits, these were merely carried through the analysis in descriptive form.

Using the new total input and output flows and revised estimates of unit values, the economist prepared a new total value flow table and estimated the revised economic rate of return as being 11.2 per cent per year.

She also revised estimates of income effects by income groups, using the results of the labour survey. She confirmed the balance of payments effect developed during the Stage I appraisal.

The major effort during the Stage II appraisal process related to the calculation of measures of financial worth of the project for the different farmer groups identified. As mentioned earlier, ten groups were identified on the basis of farm size (income) and site classes. A separate financial analysis was carried out for each group, using more detailed productivity information by site classes, information on alternative rates of return obtained from a small sample survey of members of each of the ten income/site groups, and other information on cost differences between the groups. Using the results of these analyses, the economist developed a table which showed results (FRRs) by income/site class groups (see Table 5.15). The conclusion based on this analysis was that the project had a fairly good chance of gaining farmer participation on the basis of financial considerations. However, considerations other than financial ones often motivate farmers with this type of project. Thus, the economist supported the social scientist in investigating these factors and general community attitudes towards the project and government involvement in the region.

Table 5.15. EXPECTED FINANCIAL RATE OF RETURN BY FARMER INCOME/SITE CLASS GROUPS
Per cent/year

Groups by income/site class	Financial rate of return	Alternative rate of return
1	13.5	10
2	12.0	10
3	10.0	10
4	13.5	8
5	12.5	8
6	11.0	8
7	10.0	8
8	13.0	7
9	11.0	7
10	10.0	7

The Stage II appraisal resulted in several changes. First, planting activities were phased over a three-year period. Second, the Stage I analysis was based on average figures for a "typical" farmer. In Stage II, farmers were divided up into subgroups, based on incomes (farm sizes) and site classes within income groups[7]. With the exception of Group 3 (high income/lowest site productivity), the calculated FRRs were above the alternative rates of return available to farmers.

Third, watershed benefits were added to economic value flows table and their inclusion resulted in a higher economic rate of return than the one calculated in the Stage I appraisal.

Conclusions based on the Stage II economic and financial appraisals were that the project was acceptable in terms of all of the five financial and economic questions raised initially. Thus, if the project passed Stage II appraisal in terms of social and environmental effects, it would be worthwhile to proceed with the final appraisal stage. (For the purposes of this discussion, we assume that the Stage II social and environmental assessments showed satisfactory results).

Stage III: Thorough Appraisal of the Most Promising Project Alternatives

This third appraisal stage, in terms of the financial and economic analyses, primarily involved refinement of the Stage II appraisal and preparation of the final project document.

The economist disaggregated inputs and outputs for the lower site lands and recalculated economic internal rates of return for those lands. It was found that the ERRs were still considerably above the social discount rate of 4 per cent.

Additional work was done on price assumptions, but it was found that the values developed for Stage II were acceptable estimates of future prices, given the level of uncertainty involved. Thus, they were not changed. However, the economist did calculate breakeven price levels (i.e. those prices that would just make the project have a 4 per cent ERR or a NPW of zero). It was found that these prices were well below the most pessimistic expectations for future prices.

Figure 5.2. THE PROCEDURE IN ECONOMIC AND FINANCIAL ASSESSMENT
To be followed for each project alternative

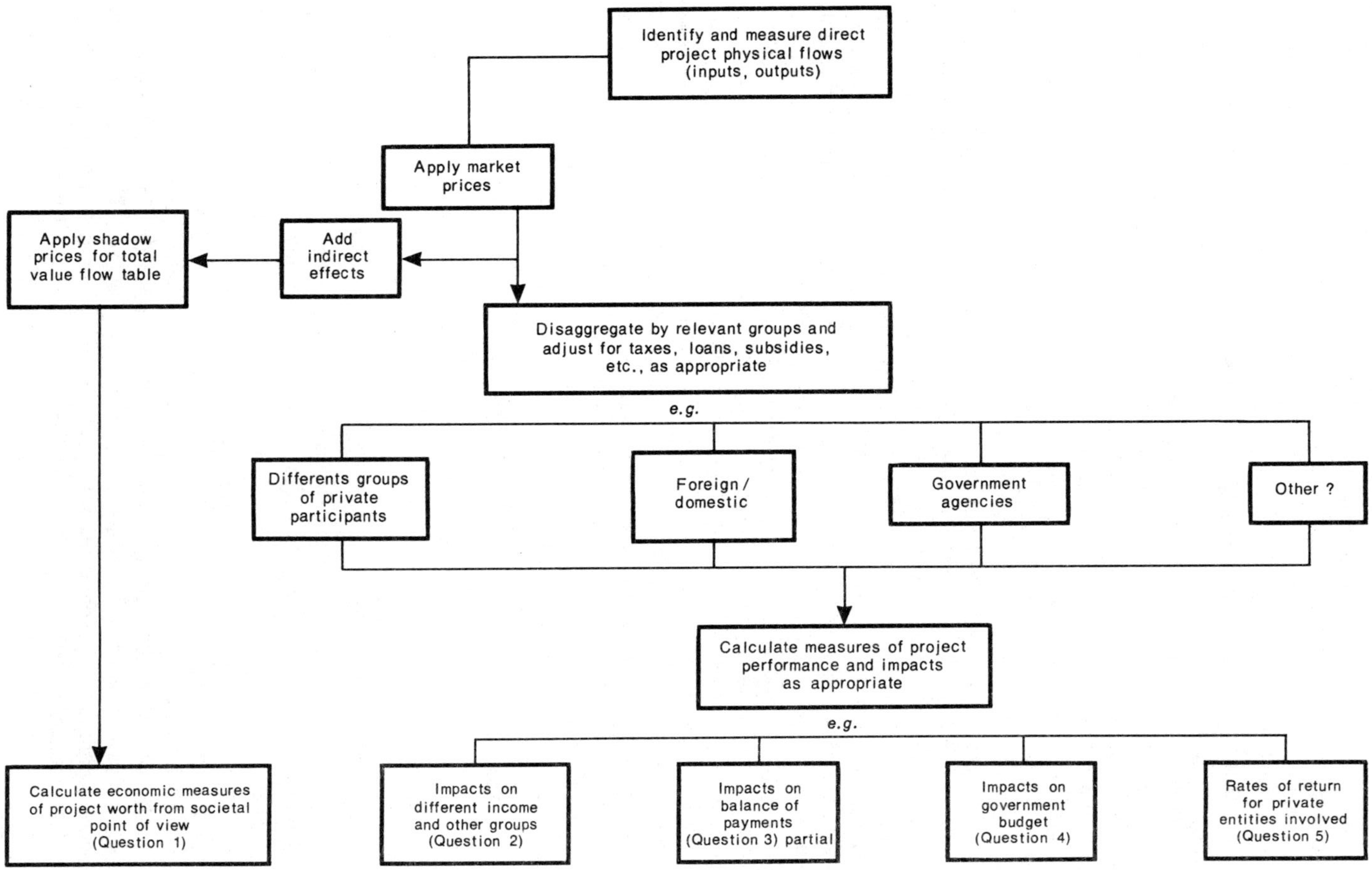

The economist did further work on estimating alternative rates of return for the ten income/site class groups included in the project analysis. Particular emphasis was placed on group 3, for which the estimated financial rate of return was just equal to the estimated alternative rate of return available to the group. It was concluded that this group would not likely participate in the project without additional subsidies.

Further work was also done on the fuelwood component of the project. Specifically, the appraisal team looked at the likely future market for fuelwood in relation to prices and availability of competing fuels. It was decided that the Stage II results were acceptable.

Stage III was finalised in a project report, including detailed discussion of social, environmental and economic and financial effects.

Note that in this particular project case, the Stage III appraisal of economic and financial effects mainly involved verification of the Stage II appraisal results. A few changes were made, but it was found that the Stage II appraisal in general provided an adequate analysis of the situation with and without the proposed project.

In many cases, and particularly for more complex types of projects, one could expect considerable change in project dimensions based on the Stage III appraisal.

The present example involved a type of project that was familiar to the forestry agency authorities. The scope for alternative designs was fairly limited, since previous experience and legal constraints restricted the number of alternative designs which needed to be analysed. In other cases, a large number of alternatives might have been carried through the rough appraisal process. Several of the alternatives would then be chosen for the Stage II appraisal, based on the combined results of the social, environmental, and economic/financial assessments. We assumed in the present example, for the sake of simplicity, that the social and environmental assessments at each stage showed no need for revision of project concept and dimensions. In reality, with the interaction of the different assessments, there would likely be more revision needed. Thus, as mentioned in Chapter I, the appraisal process tends to be rather complex and involves a great deal of give and take in an iterative process of successive approximations, until a final design is accepted by all concerned.

It is worth noting that the final appraisal would involve an analysis of uncertainty regarding various project parameters, e.g. physical input/output relationships and unit values. All three impact categories –social, environmental, and economic/financial – should be subject to uncertainty (or sensitivity) analysis.

D. Summary

The analysis of financial and economic effects of forestry projects can follow a well-defined set of steps. The first step is to determine which questions are of concern for a particular project situation. The following steps then involve:

1. developing physical flow tables for the project, including breakdowns by relevant groups – private, public and whatever other sub-groups are of interest;
2. developing unit value tables, including financial or market values and shadow prices;
3. combining physical flow information with unit value information to arrive at total value flows for the project and value flows for the relevant sub-groups; and
4. calculation of measures of project worth appropriate for answering the questions of concern.

NOTES AND REFERENCES

1. Market prices are those which are paid (received) in markets for traded goods and services. Shadow prices are estimates of real economic values in cases where market prices do not reflect a true measure of willingness to pay for goods and services or where no market prices exist. Shadow pricing is a rather complex technical topic which is discussed in detail in Chapters 5-8 of Gregersen and Contreras (1979) and the references cited therein.
2. For more details, see Annex I. The reader is also referred to Gregersen and Contreras (1979) and references cited therein for more thorough technical treatment of these measures and how to calculate them.
3. Squire and Van der Tak (1975) suggest a conceptual approach to integrating equity and efficiency impacts in a single measure. However, this approach is not sufficiently well developed in practice to be recommended here.
4. Of the secondary activity generated per unit of primary activity.
5. There is no need to discuss Tables 5.8 (economic value flows) and 5.10 (cash flows) here, since they are derived from Tables 5.1-5.7 and will vary from stage to stage only in terms of the amount of detail shown in Tables 5.1-5.7.
6. There would obviously be a significant employment effect for lower-income persons as well.
7. This disaggregation was made for several reasons. First, there was some concern that "rich" farmers were receiving subsidies which they did not need. With the disaggregations, it would be possible to eliminate richer farmers from the subsidy programme, if so desired by decision-makers. Second, the disaggregation was made in order to see if financial rates of return would be high enough on the low site lands to justify participation in the programme.

Chapter VI

A CASE STUDY: THE MYTHOPIA WATERSHED PROJECT

This is an example of how the principles and procedures outlined in this study apply. The case study is based on a real project situation, although hypothetical data are used in order to simplify the illustration[1].

A. How the Project Idea was Conceived

Mr. Cateau, the administrator for the Mythopia Dam and Reservoir No. 1, had been concerned for some time about the rapid siltation of the Mythopia reservoir, which provides water for domestic use and irrigation of some 9 500 hectares of agricultural land. The rate of siltation is much more rapid than that envisioned when the dam was built some five years ago.

The present rate of siltation is four million m^3 per year. Present reservoir capacity is down to 100 million m^3. At the present rate of siltation, it will be four years before capacity is reduced to a point where it no longer can meet expected downstream water requirements during a normal dry season. (Domestic water requirements are increasing at about 6.2 per cent per year. Irrigation requirements are fairly constant).

The problem was discussed with the Mythopia River Basin Commission, comprised of representatives from forestry, agriculture, an environmental protection agency, and local interest groups. The Commission entrusted Mr. Cateau with the task of suggesting appropriate means to alleviate the problem.

Mr. Cateau joined with representatives of the Ministry of Agriculture and Forestry to make a quick survey of the problem and its causes. It was determined that the main problem laid with poor upstream land use practices and lack of effective tree cover on certain strategic lands. Overgrazing, extension of agricultural practices onto critical watershed lands, and poor forest practices were the main causes of increased erosion and consequent siltation of the reservoir. It was decided that the Ministry of Agriculture and Forestry, through its Department of Forest Protection, would develop an appropriate project proposal and assessment for consideration by the Mythopia River Basin Commission.

Mr. Jones, a hydrologist and land use planner, and Mr. Swenson, a forest economist, were given the task of organising the initial effort and carrying out a Stage I assessment. Their report was due in two months. Using their report, the Department would design a more detailed project and undertake the Stage II assessment.

B. Organising the Stage I Assessment

Jones and Swenson organised the task of the appraisal team as follows:

1. Identify explicitly the project goals and major constraints.
2. Quantify in general terms the present land use practices and demand parameters.
3. Identify alternatives for meeting objectives, given the present situation, and quantify the main input and output elements for alternatives.
4. Carry out very rough social, economic and environmental appraisals of the alternatives.
5. Compare alternatives and make judgements.
6. Prepare Stage I appraisal report with recommendations.

What follows is a brief description of how they went about their task.

1. Project Objectives and Major Constraints

To have something concrete to work with, the appraisal team formulated the project goal: to extend the effective life of the Mythopia reservoir by reducing the rate of siltation from four million m^3 to one million m^3 per year. The project would reduce loss of downstream benefits due to extending the amount of water available for irrigated crops and domestic water use. There were no known flooding problems in the Mythopia plains, so this was not in the formulation of the problem. A further goal was to improve productivity of upstream land use.

They also identified the project constraints:

1. Laws on land use had to be followed, although some modifications in laws and regulations might be recommended for future consideration.
2. The project should not adversely affect the poorer rural inhabitants upstream who relied on watershed lands for their livelihood. Thus, any changes in land use practices which would put a burden on such individuals would have to be considered in conjunction with subsidy or other support programmes (e.g. land exchanges) which could compensate the poorer land users for income foregone.
3. In all cases where multiple use alternatives existed for improving watershed practices, such should be given priority. For example, if vegetative cover on critical slopes could provide a cash crop, this should be given preference over cover that would provide only soil stabilisation benefits. Tradeoffs would be explicitly assessed.

2. Present Land Use and Demand for Reservoir Water

The appraisal team used recent results of a land use census, together with information from the remote censing unit of the Department of Forest Protection and field checks by air to derive a general picture of present land use on the watershed. Using this and other information gathered from annual reports of agricultural extension agents and local forestry officials, the appraisal team also put together a rough assessment of land use practices and indicators of erosion rates for each.

Using data on historical trends of water consumption downstream and a quick, very extensive survey of user groups, the team estimated future trends in water use. For the first four years, use would expand with demand. After that period, use would be constrained by capacity of the reservoir to supply water. This information is shown in Table 6.1., for the "with"

Table 6.1. WATERSHED PROJECT: IDENTIFICATION OF WATER BENEFITS

Millions of m^3/year

Year	Reservoir capacity		Difference with and without project	Reservoir time		Difference in use with and without project[c]
	Without project	With project		Without project[a]	With project[b]	
0	100	100	0	86.0	86.0	0
1	96	99	3	86.4	86.4	0
2	92	98	6	86.8	86.8	0
3	88	97	9	87.2	87.2	0
4	84	96	12	84	87.7	3.7
5	80	95	15	80	88.2	8.2
6	76	94	18	76	88.7	12.7
7	72	93	21	72	89.2	17.2
8	68	92	24	68	89.8	22.8
9	64	91	27	64	90.4	26.4
10	60	90	30	60	90	30.0
11	56	89	33	56	89	33
12	52	88	36	52	88	36
13	48	87	39	48	87	39
14	44	86	42	44	86	42
15	40	85	45	40	85	45
16	36	84	48	36	84	48
17	32	83	51	32	83	51
18	28	82	54	28	82	54
19	24	81	57	24	81	57
20	20	80	60	20	80	60
21	16	79	63	16	79	63
22	12	78	66	12	78	66
23	8	77	69	8	77	69
24	4	76	72	4	76	72
25	0	75	75	0	75	75

a) Constrained by demand for water during first four years then constrained by capacity as demand outstrips supply.
b) Constrained by demand for first ten years then constrained by capacity as demand outstrips capacity even with the project.
c) This is the measure due to the project, i.e. the difference in use with and without the project.

and "without project" situations. (Note that the "with project" situation was based on the goal of reducing annual rates of siltation from four to one million m3). The figures shown are gross water use, including evaporation from the water system, estimated to be around 54 million m3 per year, with or without the project. The difference between the "with" and "without project" water use is shown in the last column and represents the basis on which water benefits would be estimated.

3. *Project Components*

Based on past experience with similar types of problems, the appraisal team developed a list of project components to be considered when analysing costs of reducing erosion and siltation.

- Establish protection forest in the most critical areas where no other activity should take place because of slope or critical nature of soil protection.
- Establish protection/production forests in areas that need permanent protection but which are less critical so that some forest utilisation can take place on a controlled basis.

- Build terraces in some of the most critical areas with very unstable soils.
- Manage and maintain pasture lands on a rotation basis, determined by their carrying capacity and ability to regenerate. This would primarily involve control and policing activities together with technical assistance.
- Establish forest management on existing natural forest areas. This would include control on harvest and other activities, watershed protection inputs into access road establishment, inventory, and other information gathering activities.
- Establish an overall watershed management and administration unit within the regional government to supervise and control implementation of an integrated watershed management programme for the whole watershed, including the above elements. Include extension services for local farmers.

a) Project Inputs (Cost Elements)

Based on the above and the information generated on land uses and land characteristics, a table was prepared showing the parts of the watershed associated with each of the components (see Table 6.2). There were very rough preliminary estimates which were needed to arrive at gross budget, manpower and other requirements for the project. More detailed breakdowns would be made, if the project passed the Stage I assessment.

Using rules of thumb developed from experience in other areas with each of the project component types, appropriate technology, labour and other input requirements and timing for each of the project components were determined. In addition, the population census figures were used to project populations in the upstream watershed areas which would be affected by the project. The population was divided by project areas into general income categories, keeping in mind the equity or income redistribution constraint mentioned earlier.

Table 6.2. AREAS ASSOCIATED WITH EACH PROJECT COMPONENT

Component	Hectares
Protection plantings	760
Protection/production plantings	870
Terraces	320
Pasture use control	3 850
Natural forest management	3 160
Watershed planning and administration	(17 500)[a]

a) Including those parts of the watershed not requiring direct action.

b) Outputs/benefits

Based on figures derived from the Ministry of Agriculture related to crop production on irrigated lands, estimates of crop losses avoided due to the additional water available (see last column, Table 6.1) were developed. Based on this analysis, it was estimated that each additional m^3 of water flowing out of the reservoir (after capacity was reached) would return a net of P2 in crop benefits[2].

In terms of increased domestic water use, the appraisal team quickly reached the conclusion that at this stage there was no effective way of developing quantitative estimates of

Table 6.3. INPUTS AND OUTPUTS: PRODUCTION FOREST COMPONENT

	0	1	2	3	4	5	6	7	8	9	10	11-16	17	18	19	20	21
Ha. planted	174	174	174	174	174								174	174	174	174	174
Thinning harvest																	
Ha.							174	174	174	174	174						
M^3/ha.							20	20	20	20	20						
Total m^3							3 480	3 480	3 480	3 480	3 480						
Final harvest																	
Ha.													174	174	174	174	174
M^3/ha.													525	525	525	525	525
Total m^3 (1000)													91.3	91.3	91.3	91.3	91.3

benefit value. In physical terms, the local downstream water board estimates indicated an annual increase in domestic water use of some 6.2 per cent per year. This increase had been included in the earlier estimates of gross water use with and without the project (see Table 6.1, last three columns).

In addition to the direct benefits associated with increased reservoir capacity, there would be some timber related benefits from the combined production/protection plantings. Based on experience elsewhere, these were expected to be as shown in Table 6.3. In years six through ten there would be some minor thinning volumes available and in years 17 through 21 there would be final harvest volumes available.

In addition to the water and timber related benefits, the following indirect benefits were identified but not quantified in the Stage I assessment:

- Eventual increases in livestock production due to regulation of grazing on watershed lands. (At the time, many of the pastures were marginal due to overgrazing). The project would restore these lands.
- Aesthetic values would increase as the land was rehabilitated.
- Access roads required for protection and other watershed management activities would permit faster and cheaper access by farmers to markets and increased mobility for extension personnel so they could reach more farmers.
- The project was expected to result in an increase in water quality in addition to quantity. A reduction in suspended loads carried over the reservoir dam would decrease the need for maintenance on individual irrigation installations.

With completion of the above, all the basic project input and output information was in hand to carry out the social, economic and environmental assessments of effects. Each of these would, of course, require some additional specific information.

4. *The Appraisal Team Gets to Work*

In this Stage I assessment, Jones and Swenson carried out the assessments themselves with occasional input from specialists in the three main areas of concern. The assessments were reviewed by the specialists for completeness and soundness of assumptions and results. Though each of the three assessments are described below separately for clarity of exposition, in fact they were carried out simultaneously by Jones and Swenson.

Asking themselves the three questions mentioned in Chapter III, the assessment team identified the following parameters as being relevant to the social assessment.

First, the team found that the project would probably have the following social effects:

- The number of skilled jobs would increase; the number of unskilled jobs particularly related to livestock activities would diminish initially, but increase over time as pasture-improvement programmes began to have positive effects on carrying capacity. Unskilled jobs in forestry would increase but be of a seasonal nature.
- Working conditions would in general improve over time as access improved and as intensity of land use increased with the improvement of land quality.
- There would be insignificant effects on population structure. No major migrations were expected.
- Community structure would likely improve with improved access to markets and increased local spending when forestry benefits increased. Stability would increase as land use practices become stabilised. (As mentioned earlier, severe overgrazing on some areas was already causing disruption as owners/users of the land found that they no longer could derive acceptable income from the land).

- The project would have an effect on local political control. Increased regional and national government involvement might cause some problems in terms of local community acceptance.
- The project would affect local cultural traditions associated with extensive land use and farmers' independence. More co-operative efforts in local communities would be needed to implement the project, particularly in the case of the production forestry activities, which most likely would have to be partially set up as co-operatives.

In answer to the second question – the intensity of effects – the appraisal team found that the temporary reduction in grazing and farm land, caused by constraints to improve land use and to reduce erosion would have the biggest impact. It would directly lessen income-generating possibilities in the short term.

Another important social effect was the disruption of the local population's sense of independence; they would have to change their habits if more co-operative activities were introduced. This might provoke resistance. Therefore, it would be wise to get them to participate in the planning and decision-making as early as possible. Citizens would have to be invited to join advisory and planning groups initiated by the project planners.

As for the third question – which groups would be most affected? – the appraisal team found that it would be the poorer members of the communities in the watershed/project region that would be most directly affected in a negative way. This would be a short-term effect caused by land use controls, but in the long term, they would stand to gain from the project. The principal beneficiaries in the short term would be the downstream water users, mainly urban dwellers and large corporate farms using irrigation. The biggest financial burden would fall on the public sector, since it was the State that would do most of the funding.

These parameters were checked with local leaders and with social scientists in the Ministry of Agriculture. (The Ministry has a small "social assessment" unit which provides such expert services to all departments and planning groups within the Ministry).

Initial discussion with various groups led to the conclusion that there would be no major negative social effects which would make the project unacceptable. Critical effects related to

Table 6.4. VALUE

Millions

	0	1	2	3	4	5	6	7	8
Benefits									
1. Irrigation and domestic use	0	0	0	0	7.4	16.4	25.4	34.4	45.6
2. Wood production							1.0	1.0	1.0
3. Total	0	0	0	0	7.4	16.4	26.4	35.4	46.6
Costs									
4. Planting protection forest	3.4	4.5	1.1						
5. Planting production forest	1.3	1.6	1.6	1.6	1.6	0.3			
6. Terrace construction	0.8								
7. Management costs[a]	1.0	1.0	1.0	1.0	1.0	1.0	1.0	1.0	1.0
8. Total	6.5	7.1	3.7	2.6	2.6	1.3	1.0	1.0	1.0
9. *Net benefit (cost)*	(6.5	7.1	3.7	2.6)	4.8	15.1	25.4	34.4	45.6
10. Present value at 12 %	(6.5	6.3	2.9	1.8)	3.0	12.9	12.9	15.6	18.4
11. NPW at 12 %	NPW at 12% P 292 million								

a) Including protection and extension services, maintenance and administration.

poorer members of the project area population could be taken care of with subsidy and self-help programmes which would be built into the project design.

To complete the Stage I social assessment, Jones and Swenson put together some summary tables indicating numbers of persons by various income levels who would be affected by the land use control components of the project and by the intended forestry activities. Interviews with local leaders and rural development specialists familiar with the project region were carried out.

Based on the results of this assessment, Jones and Swenson recommended that:

- a land use demonstration component be added to the project to help overcome local resistance to land use changes;
- the small farmer training component be expanded:
- a comprehensive incentive programme for local citizens be included to help in the transition period when land use practices would be changing;
- the State supports the establishment of small owner operated co-operatives for the forest production component. (Much of the land to be forested was in the hands of small owners.) State subsidies for planting would also be needed.

Jones and Swenson also looked at each of the relevant questions in their Stage I economic and financial assessment.

The first question was: Is the value of the benefits greater than the value of the costs of the project, when both are measured in terms of real social values (willingness to pay)? To answer this question, quantitative estimates of shadow prices were needed[3]. The value of crop losses avoided had already been calculated to be P2 per additional m^3 of water available. This was the main benefit value for which monetary estimates were available. Given the lack of information about domestic water use value, it was decided to also use this figure for the value of domestic water use. Thus, multiplying P2 times the additional water due to the project (last column, Table 6.1), they arrived at the water related benefits shown in line 1 of Table 6.4.

The wood production benefits were valued at P290 per m^3, based on import prices for wood. (Project wood would substitute for imports). The import price was adjusted down by

FLOW TABLE

of pesos

9	10	11	12	13	14	15	16	17	18	19	20	21	22	23	24	25
52.8	60	66	72	78	84	90	96	102	108	114	120	126	132	138	144	150
1.0	1.0							26	26	26	26	26				
53.8	61	66	72	78	84	90	96	128	134	140	146	152	132	138	144	150
								1.3	1.6	1.6	1.6	1.6	0.3			
1.0	1.0	1.0	1.0	1.0	1.0	1.0	1.0	1.0	1.0	1.0	1.0	1.0	1.0	1.0	1.0	1.0
1.0	1.0	1.0	1.0	1.0	1.0	1.0	1.0	2.3	2.6	2.6	2.6	2.6	1.3	1.0	1.0	1.0
52.8	60	65	71	77	83	89	95	125.4	131.4	137.4	143.4	149.4	130.7	137	143	149
19.0	19.3	18.7	18.2	17.6	17	16.3	15.5	18.3	17.1	15.9	14.9	13.8	10.8	10.1	9.4	8.8

ten per cent to reflect the lower average quality of project wood. Total wood production benefits are shown on line 2 of Table 6.4. Constant real values were assumed.

Other benefits were not valued in monetary terms due to lack of data or inappropriateness of attempting to quantify values at this stage.

Costs shown in Table 6.4 were based on market prices for machinery, equipment, structures, etc. Labour was shadow priced 30 per cent below actual minimum wages paid in the project region, since there was fairly high structural (long-term) unemployment in the region and the project would use many of the unemployed persons. Management costs (government supervision, extension personnel, etc.) were valued on the basis of standard cost factors available from the Ministry of Agriculture.

Using the Nationwide social discount rate of twelve per cent (obtained from the National Office of Management and Budget), the net present worth for the project was estimated at P292 million. The internal rate of return was estimated to be well in excess of 50 per cent. Jones and Swenson felt that such a high rate of return needed explanation, so they included the following statement in their report:

> "The high returns to this project can be explained quite easily. Since the reservoir was already in place and its cost represented 'sunk costs', they were not included in the analysis of the project. Thus, the small amount of additional expenditure required for the watershed protection activities (the project) were compared with the returns which actually include the total incremental benefits from the reservoir. Obviously, if one were analysing a new reservoir project, the situation would be quite different, since the substantial expenditure for the reservoir would have to be added into the cost stream for the project, while the benefits would remain approximately the same. The approach used is consistent with normal economic analysis procedures."

The basic information to assess the second question – What are the Distributional Impacts of the Project? – was also needed for the social impact assessment and was described earlier. Land use controls on upland areas and resulting changes in land use (see Table 6.2) would affect the poorer members of the community most directly in the short term. After a period of time, when soil quality and grazing potentials were improved, increased incomes would be possible on a sustained basis. Similarly, wood production would eventually increase the incomes of the small landowners in the project region. Based on technical considerations related to the time involved in land restoration and timber production, Jones and Swenson prepared several tables showing estimated changes in incomes, by income categories.

These tables provided estimates of gross magnitude and timing of subsidies needed in order to ensure that earnings of lower income persons would not decrease due to the project. During the first twelve years, subsidies in varying amounts would be needed. After that time, the analysis indicated that incomes would be higher with the project than without it, and subsidies could be discontinued.

Jones and Swenson suggested that additional water charges on downstream water users might help finance the subsidies needed for upstream landowners, who would have to make sacrifices to assure the downstream benefits.

The third question – How will the project affect economic stability? – included the effect on balance of payments and impacts on community stability. Jones and Swenson calculated the direct balance of payments effect to be equal to the net import savings associated with the wood production. In addition, some of the irrigated crops produced with project water would be exported. They provided a rough estimate of the net crop export value. Shadow prices for foreign exchange were not used, since the country had a free exchange rate and no balance of payments problems.

Community stability would be positively affected over the longer term, due to stabilisation of land use. In the shorter term – during the period when landowners would have to curtail grazing activities – it was considered that the State subsidies would take care of any potential economic instability problems.

The fourth question – Is the project within the budget limitations of the government agencies involved? – was not relevant here, since no budget constraint was placed on the effort. If the project met the other criteria for acceptance, funds would be made available through the Mythopia River Basin Commission. However, a financial budget estimate was made for the project, detailing the timing and magnitude of likely State financing requirements.

Question five – Is the project financially attractive to all private entities who will have to put their resources into the project – relates directly to the distributional effects discussed earlier. Upstream landowners would have no economic incentive to curtail land use practices. Earlier calculations concerned with the distributional impacts of the project indicated that there would be a net income loss in present value terms for upstream landowners who would be forced to curtail various land uses. Later benefits due to wood production and improved land use were not high enough to make up for initial losses, when all values were discounted back to the present, using an average alternative rate of return of 13 per cent (which was considered as a representative figure for the persons involved). Thus, the answer to this question was that subsidies would be needed in order to secure voluntary participation in the project. Regulations would take care of some of the problem, but much of the participation in improving land use practices to reduce erosion would have to be voluntary.

In sum, the Stage I economic and financial appraisal indicated that the economic return for the project was high, the foreign exchange and economic stability impacts were positive, and the State financing requirement did not exceed any constraint level. However, the distributional impacts of the project would be unacceptable unless subsidies were provided. Similarly, voluntary participation by private landowners would be secured only if they received subsidies. Thus, the question of an appropriate subsidy or incentive package was flagged as a critical factor to consider in the Stage II and III assessments.

As far as the environment impact was concerned, Jones and Swenson considered that the proposed project, a priori, brought nothing but positive effects as far as the environment was concerned: it made it possible to control erosion and reconstitute the soils, to regularise the water system, increase the quantity of the plant biomass produced and, Jones and Swenson believed, to improve the flora, the fauna and the quality of the landscape. So as not to miss any possible effect, they consulted a checklist used to conduct impact studies. During this review, they discovered some effects of the project that might possibly be negative: the effects of fertiliser on the quality of water and those due to herbicides and pesticides.

As Jones and Swenson were conscientious and as one was an economist and the other a water specialist, they did not feel fully competent to judge the environmental questions. They decided to call in a specialist to complete their appraisal. But they knew no one in Mythopia with in-depth knowledge of environmental effects of the type envisaged in the project. They therefore approached an international organisation which recommended a foreign expert, Mr. Dendrophilos, known for his great knowledge of forest ecosystems in climates analogous to that of Mythopia.

When he studied the project, Mr. Dendrophilos was surprised to discover two things.

The project made the basic assumption that the work envisaged would reduce the silting of the dam from four to one million m^3 per annum. Dendrophilos found this figure very optimistic and wondered whether the work would really bring about such a level of improvement. Admittedly, similar projects had already been conducted in other river basins of

the Mythopia River. However, no measuring arrangements had been prepared to make an ex post assessment of their efficiency and monitor the effective reduction of erosion. Dendrophilos consulted the world literature on the subject but found no references relating to soil types and climate such as those of Mythopia. As he did not have sufficiently sound data to calculate the effects of the project on erosion and silting, he made two proposals: on the one hand, he asked that sensitivity analyses be made using the following quantities of sediment as average annual silting assumption for the reservoir:
3.10^6 m^3/year; 2.10^6 m^3/year; $1.5.10^6$ m^3/year. On the other hand, he suggested that the project be supplemented by experimental measuring arrangements so as to collect data on efficiency of the works that would be very useful in appraising other projects bound to arise in Mythopia.

The second thing that surprised Dendrophilos was the forecasts for forest productivity: a timber harvest of 525 m^3 per hectare 17 years after planting, i.e. an average productivity of nearly 31 cubic metres per annum (not including the thinning products). Dendrophilos came from Syldavia, a country where the majority of plantation projects provided for the production of approximately eight m^3 per ha per annum but where the exploitation of existing forests did not provide more than four m^3/ha./year on average. Certain forests situated on particularly poor soils produced little more than two m^3. Considering the nature of the soils and climate of Mythopia, it seemed to him very optimistic to forecast a production of 31 m^3. A figure of 20 m^3 seemed a more reasonable forecast.

In addition, Dendrophilos feared that the cultivation of ultra rapid growth (A) species provided for in the project might exhaust the soil and in the long-term produce effects as bad as those which it was hoped to avoid. He therefore suggested that the project be modified either by lengthening the duration of the cycle for speciesA), or by planting more slowly growing species in the areas which were most exposed to erosion of the watershed. Species B) would have a slower growth (eight to ten m^3 per hectare per annum) for harvest at 50 or 100 years, according to the cycle chosen, but had two important advantages: better protection of the soil from erosion and a supply of lumber of a value five times greater than that of the industrial timber provided by (A) species.

He proposed that sensitivity analyses be made with various assumptions of the productivity of (A), (15, 20 and 25 m^3); with three cycles for (A) (17, 34 and 70 years); and with the introduction of (B) on half the reafforested area. These assumptions are given in Table 6.5.

In order to measure the effects of the different possible variables on erosion, he suggested including in the experimental arrangements measures of areas planted with (A) (with cutting

Table 6.5. SENSITIVITY ANALYSES

Assumptions	Species planted %	Duration of cycle	Productivity in m^3 (per annum)	Average annual silting in 10^6 m^3
1	100 A	17	15	2
2	100 A	17	20	2
3	100 A	17	25	2
4	100 A	34	20	1.5
5	100 A	70	20	1
6	50 A	17	20	1.5
	50 B	50	10	
7	50 A	34	20	1
	50 B	100	10	

planned at 17, 34 and 70 years) and of areas planted with (B) (with cutting at 50 and 100 years).

In addition, since he feared that the variables with long cycles would be at a disadvantage because of the effect of the discounting rate, he proposed that the sensitivity analyses be made with various assumptions about the long-term evolution of the shadow prices for timber.

Dendrophilos also suggested that new methods of agro-sylvo-pastoral production be included in the project. This suggestion derived from the following observation: the project was based on zoning which made separate arrangements for the reafforestation of certain areas and the improvement of pastoral techniques in other areas. However, he was acquainted with promising agro-sylvo-pastoral projects, particularly in Syldavia. They were based on a group of techniques for the rationalised management of livestock together with the establishment of pasturage under forest canopy, crushing the brushwood under the forests and spreading artificial fertilisers. These techniques made it possible to create grasslands under forest populations which were as productive in fodder units as directly-seeded grass, with no danger of erosion. But above all it became possible on a given area to produce both timber and livestock feed and thus considerably improve farmers' incomes. However, the development of this new method of production in Syldavia had required a considerable technical and organising staff among the farmers. Dendrophilos therefore suggested that similar efforts be undertaken in Mythopia and that, in addition to the scientific experimental arrangement designed to better understand the effects of erosion, the project should include a socio-economic experiment based on these new methods of combined sylvo-pastoral production.

Using the material developed above, Jones and Swenson prepared a brief 24-page report, highlighting their findings and recommendations. The Mythopia River Basin Comission recommended that a more detailed Stage II assessment and planning effort be undertaken, taking into account the suggested modifications derived from the Stage I analysis. The Ministry of Agriculture authorised the Forest Protection Department to carry out and co-ordinate this next stage.

C. Moving Ahead on the Stage II Assessment

Jones and Swenson were assigned the task of organising the Stage II assessment. In addition, specialists in extension, social assessment and environmental impacts were added to the team. Arrangements were made for input from specialists on subsidy programmes, livestock management, soils, and sylviculture and plantation management.

Jones and Swenson envisioned the following activities for the Stage II effort:

- Analyse in more detail land use patterns in the project area, also specifying ownership size distribution, population affected by project and other statistics on the project area needed for detailed planning of project components;
- Check Stage I estimates of downstream water use for irrigation and domestic purposes and revise as needed. Attempt to develop a better value estimate for additional domestic water use due to project.
- Carry out a more detailed assessment of social effects, given the fact that these appear to be the most critical. Pay particular attention to means for gaining local community support and participation, effects on the poorer members of the communities, and design of demonstration projects with special attention to new techniques in agro-sylvo-pastoral work;
- Undertake sensitivity analyses concerning the most uncertain elements and particularly those proposed by Dendrophilos in relation to the effects of the project on

erosion, annual growth of trees and long-term shadow prices of timber. Check areas in Table 6.2 for application of different project components. To the extent possible, develop separate sets of input-output data to facilitate analysis of each separable component of the project;

- Revise the economic and financial analyses using the more detailed project data. Carry out analyses for separable components to determine efficiency of each component;
- Analyse in more detail the probable magnitudes and timing of effects on employment and income for different income groups in project region;
- Develop the subsidy/incentive component of the project in detail, considering necessary magnitudes and timing of needed subsidies and other incentive programmes over the life of the project. Check results with local community leaders involved in the project advisory group;
- Study in more detail the environmental effects identified as being significant in the Stage I assessment, and prepare an experimental arrangement designed to measure the effects of the project on erosion.

Members of the expanded appraisal team were assigned to specific tasks according to their expertise and qualifications. Co-ordinated field surveys were organised to generate the necessary data for the assessment. Field survey plans were checked and co-ordinated with local community leaders' efforts to supply information. The field survey was carried out over a two-month period.

Using the more detailed new information, separate input-output accounts were developed for each of the major project components. Based on the economic analysis of these components, it was found that all of them could be justified separately, except for two. For the forest protection component, not enough information was available on separate benefits to carry out an economic analysis. Also, no separate economic analysis of the overall watershed management and administration unit was possible. This type of overhead activity is not separable from the other elements in terms of benefits (i.e. the benefits depend directly on the implementation of the other components).

Jones and Swenson conducted sensitivity analyses concerning an increase of ten per cent in the costs and a reduction of 20 per cent in the benefits linked with improved use of downstream water. They also carried out the sensitivity analyses proposed by Dendrophilos. These led the appraisal team to two conclusions:

It was not necessary to decide at what age the trees should be cut in the plantations involved in the project. The principle of a cutting at 17 years was maintained but the figure could subsequently be raised if the results of the experimental arrangement indicated that this would reduce erosion in a significant way.

The recommendation to plant 50 per cent of the reafforested area with rapidly growing species (A) and 50 per cent with more slow growing species (B) was retained because it would supply the lumber which the country needed and might improve the effects of erosion control.

This variable was subject to further study by the soils and plantation forestry experts. They checked alternatives for the plantation component, including origins, appropriate planting designs and need for maintenance activities. As a result of these efforts, some adjustments were made in the areas to be planted for "protection only" and those that could produce both protection and wood outputs. The latter were expanded somewhat, given the conclusions of the technical group that proper forest management for wood production could be practised on some areas which had previously been classified as "protection only". An agroforestry component was added.

Due to the high cost of terrace construction in relation to estimated benefits, the terraced area was reduced to a minimum, while protection forest activities were expanded. Of particular concern in the Stage II assessment was availability of inputs when required according to project plans. Inputs studied in detail included labour, seedlings for the plantation component, fertiliser for the pasture land improvement component, and contract services for road expansion and improvement. Due to lack of adequate seedling availability, the plantation component was phased out over a period of four years. A nursery component was added. Labour and fertiliser were found to be readily available in the quantities needed. Road construction and improvement activities would have to rely to a great extent on contractors brought in from elsewhere. Road cost estimates were revised upward to reflect higher costs for the imported contractors.

Extension personnel requirements were checked with the Stage I estimates. It was decided that the initial estimates were acceptable. The cost of adding demonstration plots for both forestry, grazing land and agro-sylvo-pastoral improvement was estimated for inclusion in the economic analysis.

Dendrophilos organised a mission to Syldavia for the appraisal team to study the agro-sylvo-pastoral system there. On their return, the sociologist of the team conducted an enquiry among the farmers of the upper watershed to determine whether some of them might be willing to undertake an analogous experiment. His survey showed that enough livestock producers were interested, provided subsidies and technical aid were granted.

The subsidy specialist and Swenson developed a detailed plan for subsidies, including some direct cost share elements together with fiscal incentives for the larger landowners and free seedlings for landowners. Detailed estimates of the impacts of grazing curtailment (in connection with the pasture improvement component) were developed. The income losses were considered to be a measure of the subsidies or payments that would have to be given to those forced to curtail their grazing activities. (At this stage, it was decided to ignore the fact that future incomes of such persons would increase due to pasture improvement. Given the government's interest in improving the conditions of the poorer families in the project region – generally the owners of the grazing animals – this was considered to be justified.) It was estimated that the total subsidy amount needed for the project would be in excess of P1 million per year for the first decade. This would be a direct financial cost to the government. It was added to the revised estimate of the total public financial cost of the project (the budget estimate).

The social effects identified in Stage I were checked more thoroughly. The social scientist found that the major social effects had been identified, as had the type of impacts on different groups. More detailed analysis was made of several critical social effects, and means for overcoming local resistance to the project and to increased regional governmental control in the project area were designed. Appropriate means for stimulating and organising landowner co-operatives for the forestry/wood production component were investigated. In the case of protection forests, the government would bear the total cost of planting, management, and technical services, as the landowners themselves would receive few benefits from such activities.

Calculations of economic effects of the project were redone using the new, more detailed data. The net present worth dropped slightly, to P 251 million. The rate of return dropped to 42 per cent, still very high by any standard.

The environmental assessment was expanded to look at effects of fertilisers on water quality and to estimate impacts due to use of herbicides and pesticides.

Lastly, Dendrophilos spelled out the experimental arrangement that he proposed to introduce to measure the effects of the work which would be undertaken on erosion.

These results and revisions in the project concept and design were reviewed and a final Stage II report was prepared, recommending that the project be accepted. No insurmountable bottlenecks were identified. The public budget estimate had increased considerably over that estimated in the Stage I assessment, and the new figure created some concern among members of the Mythopia River Basin Commission. This was flagged as a key item for consideration in the Stage III final appraisal. In addition, the problem of local participation still remained a potential bottleneck and this would also be the focus of the Stage III assessment.

The Commission recommended that a final project document, the Stage III report, be prepared:

D. The Stage III Assessment: Preparing the Final Report

Since the project would have be approved and financed by the Mythopia River Basin Commission (MRBC) with a special budget allocation from the national government, the Commission took the lead at this stage. Mr. Leuschner, a project planner from the MRBC was assigned the task of preparing the final assessment. He was supported by Jones and Swenson, Dendrophilos and other experts.

Based on review of the Stage I and II documents, it was decided that the Stage III assessment would concentrate on the following:

- Check figures and calculations presented in the Stage II report for accuracy and completeness;
- Assess means for reducing the public financial burden (financial assessment);
- Design detailed public involvement plan and assess it in terms of community response and local leadership ability (related to social assessment);
- Clarify mechanisms for establishment of forestry co-operatives;
- Prepare a final draft of the environmental impact assessment for public review (as required by law);
- Design monitoring and control system for the project and formal procedures for project modifications as needed (based on annual review of performance);
- Prepare final project report and budget estimate; prepare documentation necessary for formal consideration and decision.

Checking the Stage II Results

The assessment team developed systematic means for sample checks of results. Uncontrollable uncertainty surrounded certain estimates, e.g. the speed of silting of the reservoir after the works, the growth rate of the trees and the average annual water flow. (This latter point was a problem even when the dam was first contructed). There was no way to correct for the deficiency in data on all these points. Mr. Leuschner satisfied himself that the assessment team had adequately considered reasonable variations in the figures used in the sensitivity analysis.

Mr. Leuschner was not satisfied with the information provided in the Stage II report related to potential social effects. But he had already pointed this out as an area of concentration for the Stage III assessment.

In sum, the quantitative input and output and the value estimates of the Stage II report were accepted and used in the Stage III effort. An additional sensitivity analysis was made, taking low shadow prices for the timber. It was found that the rate of return on this project component would still be favourable, even if wood prices turned out to be 15 per cent lower than initially estimated.

1. Means of Reducing Public Financial Burden

The high cost of the project to the public sector had been pinpointed earlier as a major concern. Since the Stage II cost estimates had been checked and accepted as being the best possible under the given circumstances of uncertainty and available technology, the only way to lighten the public burden would be to reduce the scale of the project or to shift some of the financial cost to beneficiaries of the project. The latter course was chosen, since it was felt that the project was already as small as possible for the purpose at hand. (There were some scale economies involved. Incremental benefits for the last addition to the project were considerably in excess of incremental costs, based on the analyses of separable components).

As a result of these decisions, Mr. Leuschner called on Mr. Cateau and the local water boards to work out a formula for increasing water charges downstream. The increased revenues would be plowed back into the project to help finance it. A separate report was prepared on the options available and on the estimated changes in water demand. It was found that the price increases proposed would have little effect on the quantity demanded. This was noted in the main project report, where reference was made to the water board studies and rate increase proposals. Mr. Leuschner considered that the increased revenue brought the direct public budget down to a reasonable level.

A plan for public participation, which drew upon the experience of the social scientist on the assessment team, local leaders and agricultural extension agents, was prepared. Several meetings were held to present the project to the local communities. Their reaction was now generally favourable, though landowners wanted a greater say in land use controls.

It was suggested that local "watchdog" committees be set up to take the place to a large extent of the policing authority. They would be elected and would be responsible for policing local compliance with the changes in land use. Public officials would still have the right to spot-check compliance and the effectiveness of the local committees, but their presence in the area would be greatly reduced and therefore considerably less objectionable. This arrangement also insured direct local participation in project management.

More local participation was also proposed in decisions on subsidy allocations. This, it was felt, would make for a more sensitive system that would insure more equitable treatment for all the landowners. Local citizens would be involved in determining the subsidy-allocation formulas to be used in their areas, even though the government maintained the right to have the final say on total allocations to communities.

Regular community meetings would be held to get community feedback on implementation problems as well as suggestions for modifications.

All these suggestions were discussed at the public meetings. Also highlighted was the landowners' continuing concern about the nature and functioning of the proposed forestry co-operations.

It was determined that the project region would have some problems in marketing the wood output from the project, unless substantial support from the government were forthcoming. While the proposed co-operative structure was considered adequate for plantation and management as well as distribution of profits, there was some concern as to how such co-operatives could function effectively in the marketing area. Given the location and the structure of roundwood markets, some fear was expressed that the co-operatives in fact would be at a disadvantage and would not be able to obtain fair payment for wood output. It was proposed that additional technical forestry staff be added to the project to work directly with the co-operatives on the marketing function. Since the first outputs from the plantations would not be forthcoming until the sixth year of the project, it was considered that there was ample time to develop an appropriate market structure and wood sale contract system. Further, it was agreed that management and output from the public natural forests in the project region would be co-ordinated with the activities of the forestry co-operatives to ensure some stability in wood supply and employment in forestry activities.

A final assessment was prepared according to regulations for implementation of the National Environmental Protection Law.

2. *Project Monitoring and Control*

Using the input developed from the Stage I through III assessments, a monitoring and control system for the project was designed to take into account all the critical elements identified. The monitoring system would include both public and local citizen input. Formal means for project modifications were established, recognizing that this type of project seldom is or can be implemented as initially designed.

The follow-up and monitoring system prepared by the appraisal team comprised two particularly important aspects:

- A technical aspect corresponding to the measurement arrangement planned by Dendrophilos. This arrangement would make it possible to collect more precise data concerning the effects of the different operations proposed (reafforestation, construction of terracing, improved management of pastures, better forest management, innovations in agro-sylvo-pastoral techniques, etc.); on the environmental constituents (effects on erosion, on the average output from the watershed, on flood and low water levels, on the composition on the flora and fauna, etc.);
- A socio-economic aspect to permit an appraisal of the impact of the new pastoral and agro-forestry techniques, both in relation to each farm and in relation to the general economics of the small region constituted by the upper watershed.

The appraisal team organised the presentation of the data provided by the follow-up and monitoring system in a simple, clear and concise form so that these data could be made available to the population concerned and particularly to the supervisory committees. These data would then form what Dendrophilos called "heritage accounts" permitting improved "heritage management", i.e. improved long-term management of the territory by the population itself.

Based on the additional work done in Stage III, a final project document was prepared, including a budget for the public sector based on accounting practices used by the Office of Management and Budget of the national government. The final 45-page report and technical appendices were transmitted to the MRBC for formal consideration.

NOTES AND REFERENCES

1. Based in part on Hans Gregersen and K. Brooks, "Economic Analysis of Watershed Projects" in *FAO Forestry Paper* No. 17, Sup. 2. Economic Analysis of Forestry Projects; Readings, 1980, FAO, Rome. Note that many forestry projects in OECD Member countries interface directly with other natural resource concerns, in this agriculture and water resource development. The issue of interagency co-ordination is critical in many cases.
2. This is a net value in the sense that all costs back to the reservoir were subtracted from the additional crop value to arrive at the benefit due to the water. These costs include farming, marketing, water distribution costs, etc., but exclude any sunk costs in structures and equipment which would be incurred with or without the project. The unit of national currency is the peso (P).
3. See Chapter V.

Appendix I

TECHNICAL CONSIDERATIONS: FINANCIAL AND ECONOMIC EFFECTS

The purpose of this appendix is to explain in more technical terms some of the concepts described in Chapter V.

As we mentioned earlier, economic and financial appraisals generally involve calculations of net present worth and internal rate of return measures. Each provides information which can be used to judge whether a project brings an economically efficient use of resources. Each measure also provides additional information which is useful in comparing or ranking project alternatives.

Here we shall define the measures and show how they are calculated. We will also compare them and indicate under what circumstances each is useful.

1. Net Present Worth (NPW)[1]

NPW is a measure of the present (discounted) value of the net benefit flow of a project. It is the sum of the present value of all benefits minus the sum of the present value of all costs or the sum of present values of benefits costs for each period *t*.

More formally, it can be written as follows:

$$NPW = \sum_{t=0}^{n} \frac{B_t}{(1+i)^t} - \sum_{t=0}^{n} \frac{C_t}{(1+i)^t}$$

or

$$NPW = \sum_{t=0}^{n} \frac{B_t - C_t}{(1+i)^t}$$

where

B_t = benefits in each year t

C_t = costs in each year t

n = number of years to end of project

i = discount rate

To calculate the *NPW* measure for a project, we need estimates of the cost and benefit flows over time or the net flow $(B_t - C_t)$ as shown on the bottom line of Table 5.8 Chapter V. We also need an appropriate discount rate, i.

1.1. The Discount Rate

For private entities, the relevant discount rate depends on whether the entity is borrowing funds or using its own funds.

If the funds are borrowed, the relevant rate is the cost of capital (the interest rate paid on borrowed funds). In some cases, this rate is increased to reflect risk and uncertainty associated with a project.

When the entity's own funds are being used, the relevant rate is the highest rate of return which could be obtained by investing the entity's funds in the next best alternative to the project being appraised. Thus, it is often called the "alternative rate of return" *(ARR)*. Again some adjustment may be made, upwards or downwards, to reflect the degrees of risk and uncertainty associated with different alternatives. For example, the next best alternative may be to put the funds in a bank, which is a fairly safe investment. The particular project being appraised may be fairly risky, in which case the interest rate might be raised slightly to reflect this greater uncertainty.

In cases where both borrowed and own capital will be used, the discount rate can be a weighted average of the borrowing rate and the alternative rate of return. In cases where the alternative rate of return is above the borrowing rate and the amount which can be borrowed is limited, then the alternative rate of return is the relevant one to use, even if funds will be borrowed to undertake the project.

When the public sector is undertaking a project for social purposes, the relevant discount rate to use is the rate generally specified by government for appraisals of economic worth of public projects. This rate should reflect the social time preference of society, or the preference for present over future consumption. It is a difficult matter to estimate such future consumption. Fortunately, forestry project appraisers will seldom have to determine the rate, since it most often is established by general government policy and all agencies are required by law or administrative regulation to use that rate.

1.2. Calculating the Net Present Worth (NPW)

Given the cost and benefit streams $(B_t - C_t)$ for a project, and the relevant discount rate, it is a straightforward and simple matter to calculate the *NPW* for the project. $B - C$ for each year t is discounted back to the present, by dividing $B_t - C_t$ by $(1 + i)^t$, where i is the relevant discount rate. Summing up these discounted values (as shown in the *NPW* formula), we arrive at the measure of *NPW* for the project. In the case of the example used in Chapter V, the *NPW* associated with Table 5.8. was \$887 000.

What does this figure mean? Simply stated, it means that the project will return a net surplus of \$887 000 of consumption benefits, expressed in present value terms, given the social discount rate of four per cent assumed in the example. This tells us that the project is efficient in economic terms.

2. Internal Rate of Return (IRR)

The internal rate of return *(IRR)* can be defined as:

- the rate of discount which makes the present value of costs equal the present value of the benefits, or $NPW = 0$;
- the average earning power of invested funds or the average rate of return on the invested funds outstanding per period while they are invested in the project.

IRR is that rate of discount such that:

$$\sum_{t=0}^{n} \frac{B_t}{(1 + IRR)^t} = \sum_{t=0}^{n} \frac{C_t}{(1 + IRR)^t}$$

or

$$\sum_{t=0}^{n} \frac{(B_t - C_t)}{(1 + IRR)^t} = 0$$

where terms are as defined earlier.

2.1. *Economic Versus Financial Rates of Return*

In the case of social economic analyses, the *IRR* is referred to as the "economic rate of return" *(ERR)*, while in financial analyses, it is referred to as the "financial rate of return" *(FRR)*. Both are calculated in exactly the same fashion.

The *FRR* shows the investor what the average earning power is associated with a given investment of his funds. More specifically, it is the average rate of return on the invested funds outstanding per period while they are invested in the project.

Thus, a *FRR* of ten per cent indicates to the investor that he will receive $10 back per year for each $1 invested during the years in which the investment is left in the project. This is a useful measure for an investor, since it provides a clear means for comparing alternative uses of his funds. Say that his best use of funds, other than putting them in the project, is to put them in the bank at six per cent interest per year. He compares the rate of return on the project (ten per cent) with the rate of return from the bank (six per cent), which is called the alternative rate of return *(ARR)*, or his opportunity cost of capital, and he then knows that the project use will give a greater return than the best alternative use.

The *ERR* is similarly interpreted, except it shows the decision-maker what society can expect to receive back in consumption benefits for a given investment of its scarce resources. In other words, if he calculated *ERR* is ten per cent, this tells the decision-maker that the average annual return of consumption benefits on resources outstanding per period while they are invested in the project will be $10 for every $100 of resources invested and left in the project. The *ERR* will be compared with the social discount rate to see if the project earns enough to make it worthwhile to invest (forego consumption now in favour of future consumption). Say that the relevant social discount rate is five per cent. This means that society wants to get at least a five per cent rate of return on investment of its resources to make it worthwhile to forego present consumption in favour of investment and future consumption. If the *ERR* turns out to be ten per cent for a given project, this means that, on the average, society will get more than the minimum acceptable five per cent back. Thus, the project is economically efficient in terms of its use of scarce inputs assuming that the other two conditions for economic efficiency are met.

2.2. *Calculating IRRs*

This is no formula for calculating the *IRR* when more than one cost and/or benefit is involved. Therefore, a trial and error technique has to be used. The approach is as follows:

a) First calculate a *NPW* using a rate which is estimated to be in the neighbourhood of the expected *IRR*. If the *NPW* is negative, then the *ERR* must be lower than the discount rate used. If the *NPW* is positive, then the *IRR* must be higher than the discount rate adopted.

b) If the first *NPW* calculated is negative, then reduce the discount rate up to a point where the calculated *NPW* is positive and vice versa if the first *NPW* calculated is positive. The *IRR* must now lie between the two rates of discount used in generating the positive and negative values of *NPW*.

c) Estimate the *IRR* by using the following formula:

$$IRR = \begin{matrix}\text{Low}\\ \text{rate of}\\ \text{discount}\end{matrix} + \begin{matrix}\text{Difference}\\ \text{between both}\\ \text{rates of discount}\end{matrix} \times \frac{\text{Positive } NPW}{\text{absolute difference between positive + negative } NPWs}$$

d) Repeat steps *(a)* – *(c)* for a more precise result, if needed.

If a computer is available, or if the appraiser has access to one of the desktop calculators with capacity to calculate *IRR's*, then the process is simplified.

In the example used in Chapter V, the *ERR* and *FRR* were calculated to be 8.7 per cent and 10.9 per cent, respectively, using the figures provided on the bottom lines of Tables 5.8 and 5.10). A desktop calculator (Texas Instruments TI Programmable 59) was used.

3. Comparing IRR and NPW Measures[2]

From the above definitions, the following relationship holds: When *NPW* = zero, then the *IRR* = the discount rate used in calculating the *NPW*[3]. Given the definitions and the above relationship between the two measures, what can be said about the information provided by each of them in terms of economic efficiency?

Both measures provide information related to whether present value *(PV)* if benefit is less than, equal to, or they both provide exactly the same answer to the question of whether or not a project or project component is economically efficient. If a project is accepted as being efficient in terms of one measure (i.e. $NPW \geqslant 0$), it will also be acceptable in terms of the other measure (i.e. *ERR* ⩾ social discount rate) and vice versa.

It can be seen that either of the two measures could be used equally well to determine whether a project is economically efficient (assuming there is no less expensive way to achieve the project objectives). Thus, the choice of which of the two to calculate and use is unimportant in terms of this basic question, although the analyst obviously has to calculate the measure commonly used by the institution for which he is carrying out the analysis.

Each of the two measures provides additional information that the other does not provide. The *NPW* measure, in contrast to the *IRR*, provides information on the absolute value or magnitude of the present value of net benefits of a project. Yet it tells nothing about how large the cost will be to achieve the *NPW*. Thus, there could be a project with a *NPW* of $1 000 which costs $2 million or one with the same *NPW* that costs $5 000. Both would have the same *NPW*. On the other hand, the *IRR* is a relative measure of project worth, which gives information on the returns per unit of cost and thus provides more relevant information for comparing the benefits which can be expected from alternative uses of a limited budget. Therefore, it is more useful for ranking independent project alternatives when it is not possible, for budgetary or other reasons, to undertake all projects that meet the basic economic efficiency conditions.

This latter point brings up an important consideration. Even though a project alternative is shown to be economically efficient, using either of the two measures of economic efficiency or worth, this does not mean that the project will be undertaken. Budget considerations as well as many other factors will enter the picture and the decision-making process. If a number of alternative uses (projects) exist for a limited budget, some system of ranking projects in terms of their economic worth has to be developed to provide guidance in choosing from among the alternatives that meet the basic efficiency conditions. The two measures of project worth do not necessarily provide the same order of ranking for projects. In point of fact, the *NPW* measure provides no guidance for ranking projects that are not mutually exclusive[4].

The whole problem of choosing from among alternative uses of a given budget is complex and goes beyond the confines of an economic analysis of a given project alternative. The choice of a ranking system is often made on the basis of political objectives. Even in terms of the economic efficiency objective, one or the other of the two measures is often chosen for ideological reasons or because it is "simpler to understand".

While consideration of choices among projects for a limited budget, i.e. decision-making, is beyond the scope of this note, the following points should be noted:

i) Both measures of project worth provide the same answer to the question of whether or not a given project alternative or project component is economically efficient in and of itself (assuming that the third condition is met).

ii) Since the *NPW* provides an absolute measure of project worth, while the *IRR* only provides a relative measure of average expected returns per unit of cost (and no information on the absolute size of the net benefits), it is recommended that the analyst calculates both the *NPW* and the *IRR* for projects. Regardless of which measures will be used by decision-makers in making project decisions, information on both absolute and relative economic efficiency is useful.

iii) Finally, while calculation of the *NPW* requires that the analyst have an estimate of the appropriate discount rate in hand, the discount rate is not required in order to calculate the *IRR*. Still, in order to make use of a calculated *IRR*, i.e. in order to determine whether or not the project (or project component) being analysed represents an economically efficient use of resources, some estimate of the discount rate (or *ARR*) is needed, since the *IRR* only has meaning in the context of the other possible uses for resources.

4. Appropriate Use of NPW and IRR Measures

The basic guideline is that *IRRs* can be used in a capital budgeting exercise to rank project alternatives that are not mutually exclusive. *NPW* can best be used to compare mutually exclusive project alternatives, as explained below.

4.1. Choosing Among Mutually Exclusive Project Alternatives

If only one alternative can be chosen from several being considered, then we are dealing with "mutually exclusive" alternatives. A few examples are:

- For a given forest stand, only one rotation or cutting cycle can be chosen at a given time (if you choose to harvest in year *x* then you cannot also harvest the stand in year *y*).
- If you choose technology *M* for a given project, then you cannot also choose technology *N* for the same project.
- If you choose location *A* for a given project to meet a given requirement, than you cannot also choose location *B* to meet the same requirement.
- If you choose scale *K* for a given project, then you cannot also choose scale *L* for the same project.

The question here is, how do you choose between mutually exclusive alternatives?

The basic procedure is to rank mutually exclusive alternatives on the basis of their *NPWs* and then choose that alternative with the highest *NPW*. Ranking on the basis of *IRRs* and then choosing that alternative with the highest *IRR* can lead to the wrong decision, depending on the context of the overall problem being analysed. For example, assume that we are considering two alternative uses for a given place of forest land:

Alternative A: By investing $100 in year 0 in some intensive silvicultural activities on ten acres of land, we could expect to increase the value of the timber output in year ten from $300 to $600, i.e. we would get a return due to the project of $300 in year ten.

The *IRR* for this project (assuming that there are no other costs or returns due to the project) would be equal to 11.6 per cent. The *NPW* at six per cent (assumed to be the relevant *ARR)* is $67.

Alternative B: As an alternative, you could invest $1 per acre ($10 for the ten acres) in chemical spraying of the stand and expect to increase the value of the harvest in year ten from $300 to $350. In this case, the *IRR* would be 17.5 per cent, but the *NPW* would only be $17.90.

Both would be profitable alternatives. If the *IRRs* were used to rank project alternatives for the forest stand, alternative *B* would be chosen. But, obviously, this would not be the best use for the land, since you could earn a great deal more in absolute terms by choosing alternative *A*. If we ranked the alternatives on the basis of their *NPWs* then the right choice would be made. In *PV* terms, you would earn $67 – $17.90 = $49.10 more with alternative *A*. In future value terms (i.e. when you harvest), you would earn $300 – $50 or $250 more.

NOTES AND REFERENCES

1. Also referred to as "present net worth" or "net present value".
2. Adapted from Gregersen and Contreras (1979).
3. If *NPW* > *0*, then *IRR* > *i* used;
   If *NPW* = *0*, then *IRR* = *i* used;
   If *NPW* < *0*, then *IRR* < *i* used.
   Where *"i"* equals the discount rate used.
4. Mutually exclusive project alternatives are those for which only one of a set of alternatives can be undertaken. For example, two projects which envisage using the same area of forest land are mutually exclusive alternatives. Only one of the two can be undertaken at a given time.

Appendix II

TAKING ACCOUNT OF RENEWABLE NATURAL RESOURCES IN ECONOMIC CALCULATIONS

Man uses renewable natural resources in many of his economic activities: this is the case whenever biological processes (whether spontaneous or controlled) are involved in an important way in the process of economic production. Forest activities are typically based on exploitation of the renewable natural resources which constitute forest ecosystems. Similarly, agriculture relies on the resources constituted by the fertility potential of soils and the genetic heritage of domestic animal and plant species. Hunting and fishing are based on those which constitute the animal species populations, etc. Renewability of these resources cannot be maintained unless the conditions are respected which permit the evolution of these biological processes. Among these conditions, there is the necessity of taking account of the duration of biological cycles, particularly those of growth and reproduction among living beings. These cycles are often long, especially in the case of forestry: few forest plantations are ready for harvest in less than 80 or 100 years. Some have reproduction cycles of even longer duration (oak is around 200 years).

The length of these periods provides the explanation for a criticism which is sometimes formulated in relation to the use of discounting rates in economic calculations: this method of calculation would render it impossible to take account of the conditions of renewal of natural resources when biological cycles of several decades are involved. For example, if in order to appraise the consequences of a project use is made of a rather high discounting rate, all the consequences which take effect only in the long or very long term are entirely blotted out. A discounting rate of around ten per cent assigns to one franc, in today's terms, a value of two francs in seven years or four francs in 14 years or ... 2^n francs in seven times n years, i.e. a thousand francs in 70 years, and so on.

This appendix attempts to show how to combine the concepts which are in common use in economic calculation, namely the discounting rate and shadow prices, in order to take account of the very long-term consequences of projects relating to the management of renewable natural resources.

The principles adopted in this model are those of Claude Henry, Director of the Laboratoire d'Econométrie of the Ecole Polytechnique in Paris.

In this schematic model, we shall consider an economy revolving around two goods and extending over several successive periods $t_1, t_2, ..., tj$.

Good X is manufactured according to a technical process which does not involve any natural resource (or involves resources available in unlimited quantity). Good S, on the other hand, is produced by drawing upon a stock of natural resources which are renewed according to biological processes which human technology cannot basically accelerate.

Let us use Wj to denote the stock of good X in existence at the beginning of period t_j.

Let us call the part of this good consumed during this period x_j.

And the part of this good used by the production process during period t_j, y_j so as to rebuild a stock W_{j+1} available at the beginning of the following period. We shall represent the production process by the linear production function:

$W_{j+1} \quad = \quad (1 + a)y_j$, where a is a coefficient expressing the productivity of this process.

At time t_j, we have the equilibrium equation

(1) $W_j - x_j - y_j = 0$

At time t_{j+1}, we have the equilibrium equation

$W_{j+1} - x_{j+1} - y_{j+1} = 0$, from which we arrive, by substitution, at

(2) $(1 + a)y_j - x_{j+1} - y_{j+1} = 0$, another multiperiod form of the production function.

Here we have given the elementary bases of a very classical growth model.

Let us now consider the renewable natural resource S. Let us denote the stock available at the beginning of period t_j by S_j, the quantity consumed during this period by s_j, and the quantity remaining available for the biological processes of the resource reproduction by z_j.

Let us represent these processes by the production function $S_{j+1} = g(z_j)$.

The equilibrium equation at time t_j is written:

(3) $S_j - s_j - z_j = 0$ and, at time t_{j+1}:

$S_{j+1} - s_{j+1} - z_{j+1} = 0$, or, by substituting $g(z_j)$ for S_{j+1},

(4) $g(z_j) - s_{j+1} - z_{j+1} = 0$

In order to specify the form of function g, let us assume – which approximates to many biological processes – that function g is represented by the curve G below (z on X-axis and $g(z)$ on Y-axis).

The curve G intersects the first bisector at points A and B. C is the point where the tangent is parallel to the first bisector. M is the cursor. C and M are projected onto D and N on the first bisector parallel to the Y-axis.

Figure 1

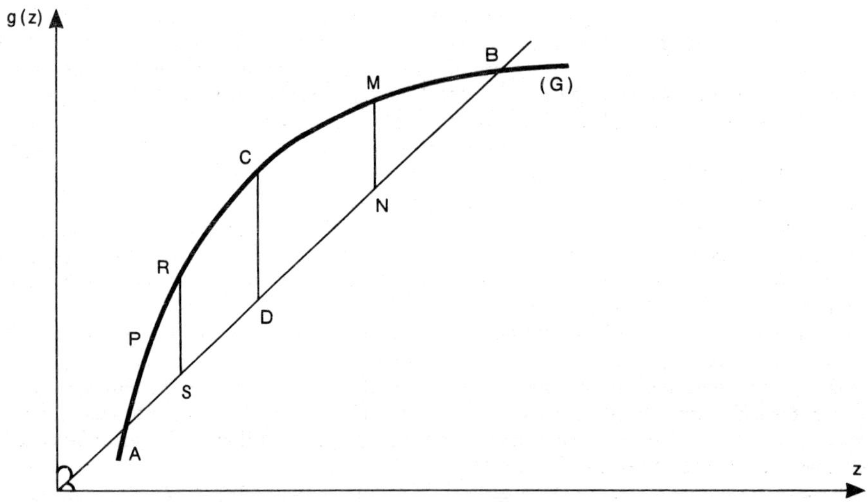

The length of the segment MN is equal to $g(z) - z$. According to equation[3], it therefore represents consumption s of good S, in a stationary system.

CD is the maximum of this length, this value being attained when the tangent is parallel to the first bisector and therefore when $g'(z) = 1$. Point B corresponds to a stable condition of the ecosystem without any withdrawal.

Beyond point B, there would only be a stable system with negative withdrawal, i.e., more of good s instead of consumption: in practice, therefore, it is not possible to exceed B, which is the point of equilibrium of the undisturbed natural environment.

Point A also corresponds to a position where withdrawal is nil, but any equilibrium at this point is unstable. It is a cut-off point: to the left of A, equilibrium is only possible with negative consumption, i.e. an addition of resources to rebuild the stock which has fallen too low for the biological processes to reconstitute it. Unless this is done, the resource will eventually become exhausted and the deterioration will be irreversible.

In conclusion, therefore, the resource can be exploited in a stable manner only between points A and B, with a maximum reached at C.

Having described the basis of our model, we shall proceed to apply it in order to draw conclusions on the use of the discounting rate and shadow prices.

We have denoted consumption series for the various successive periods $t_1, t_2 ... t_j ...$ by $(x_1, s_1)(x_2, s_2)...(x_j, s_j)...$

Let us assume that there is a collective preference function U relating to the distribution of the consumption of the two goods between the different periods:

$$U(x_1, s_1, x_2, s_2, ... x_j, s_j ...).$$

With regard to U, we shall assume only that it exists and that it can be derived. To optimise economic production, it is therefore necessary to solve the following problem: find a production programme (i.e. a series of values for the variables y_j and z_j) which maximises U subject to the constraints:

....		
C_j :	$(1 + a)y_{j-1} - x_j - y_j = 0$	p_j
C_{j+1}	$(1+a)y_j - x_{j+1} - y_{j+1} = 0$	p_{j+1}
....		
K_j :	$g(z_{j-1}) - s_j - z_j = 0$	q_j
K_{j+1} :	$g(z_j) - s_{j+1} - z_{j+1} = 0$	q_{j+1}
....		

$P_j, P_{j+1}, q_j, q_{j+1}$ denote the dual variables associated with the constraints $C_j, C_{j+1}, K_j, K_{j+1}$.

Using the Lagrange multiplier to solve this problem, gives the following:

maximise $L = U + \Sigma_j (p_j C_j + q_j K_j)$

Now, at the optimum, the partial derivatives of L in relation to all the variables y_j, z_j, x_j and s_j must equal zero, from which we get the equations:

(5) $\dfrac{dL}{dy_j} = 0$

(6) $\dfrac{dL}{dz_j} = 0$

(7) $\dfrac{dL}{dx_j} = 0$

(8) $\dfrac{dL}{ds_j} = 0$

Since these partial derivatives are written:

$$\frac{dL}{dy_j} = -p_j + (1 + a)p_{j+1}$$

$$\frac{dL}{dz_j} = -q_j + g'(z_j)q_{j+1}$$

$$\frac{dL}{dx_j} = \frac{dU}{dx_j} - p_j$$

$$\frac{dL}{ds_j} = \frac{dU}{ds_j} - q_j$$

equations (5) to (8) make it possible to describe the conditions for the economic optimum in the following form:

$$(5) \quad \frac{p_j}{p_{j+1}} = 1 + a$$

$$(6) \quad \frac{q_j}{q_{j+1}} = g'(z_j)$$

$$(7) \quad p_j = \frac{dU}{dx_j}$$

$$(8) \quad q_j = \frac{dU}{ds_j}$$

In equations (7) and (8):

p_j is the discounted shadow price of good X at time t_j (if its consumption x_j varies by a quantity dx_j, collective satisfaction measured today varies by $dU = pjdx$, and q_j is the discounted shadow price of the renewable resource S at time t_j).

Condition (5) shows that the ratio between the discounted shadow prices, p_j to p_j+1, for two successive periods, must be equal to $1 + a$ if the optimum is to be reached.

However, *"a"* seems identical with what, by definition, is called the discounting rate. In this model this cannot be fixed arbitrarily but at the optimum it depends on the coefficient $(1 + a)$ used to represent the production function of the economy.[1]

Condition (6) shows that in order to reach the economic optimum the ratio between the discounted prices, q_j to q_j+1, of the resource S must be equal to $g'(z)$.

Let us denote the shadow price of good S in relation to good X by

$$Q_j = \frac{q_j}{p_j}$$

Condition (6) can be rewritten as (9)

$$\frac{Q_j}{Q_{j+1}} = \frac{g'(z)}{1 + a}$$

Let us now consider three alternative policies. The first consists in deciding to exploit the maximum possible amount of the natural resource. This is achieved at point C of curve G, where $g'(z) = 1$

Condition (6) is written: $q_j = q_{j+1}$, which implies that the discounted shadow price S must be constant.

As for condition (9), this is written:

$Q_{j+1} = (1 + a)Q_j$, in other words, in order to exploit the maximum of a renewable natural resource in conditions of economic optimum, it must be given a shadow price which increases over time at a rate equal to the discounting rate.

A second policy would consist in deciding to keep a constant reference price for resource S, i.e. to impose the constraint $Q_{j+1} = Q_j$.

The optimum condition (9) is written $g'(z) = 1 + a > 1$.

Let us denote the point on curve G (figure 1) whose tangent has the slope $(1 + a)$ by P.

If point P is to the left of A, there is an irreversible deterioration of the resource which can no longer reconstitute itself. If P lies between C and A, the resource will be exploited without being exhausted, but only a quantity less than C will be consumed. Furthermore, the higher the coefficient a, the closer we get to the cut-off point A and therefore the greater becomes the risk of irreversibility.

Lastly, a third, intermediate policy could be to assign to S shadow prices that increase at a rate, which is lower than a; therefore with

$$\frac{Q_{j+1}}{Q_j} = 1 + \alpha$$

equation (9) becomes:

$$g'(z) = \frac{1 + a}{1 + \alpha}$$

which determines point R on curve G. The length of the segment RS corresponds to the quantity of the renewable natural resource that will be consumed at the optimum when this third policy is applied, which will be all the safer the nearer one is to C and therefore the nearer α is to a.

Thus, in order to exploit a renewable natural resource, several alternative pricing policies are possible, giving varying levels of consumption and involving varying degrees of risk of irreversibility. On the other hand, for a non-renewable resource, the range of policy options is much narrower. If we look at the last example we have, by definition:

$w_j = s_j + w_{j+1}$ because the resource is not renewable, i.e. $g(z) = z$, therefore $g'(z) = 1$.

Conditions (6) and (9) are then written:

$q_{j+1} = q_j$ and $Q_{j+1} = (1 + a)Q_j$

The only economically optimal policy consists in giving a non-renewable resource constant discounted shadow prices, i.e. current shadow prices increasing at the pace of the discounting rate.

Finally, let us examine the question which we raised at the beginning: what happens to a renewable resource whose renewal cycle covers a very long period?

Up to now the duration of the renewal of the resource has not been taken into account explicitly in our model. In fact it is reflected by the form of curve G. Let us examine figure 2.

Angle ß has the tangent $(1 + a)$.

Curve G_n relating to a non-renewable resource coincides with the first bisector.

Curve G_i relating to an unlimited resource (or a resource which renews itself immediately) coincides with the y axis.

Between these two extremes, $G1$, $G2$ and $G3$ correspond to resources which renew themselves all the more easily and rapidly when they are farthest away from the first bisector.

Curve $G3$ is farthest from it.

A first policy (constant discounting rate) permits very big withdrawals (because segment C_3D_3 is very long). The slope $g'(z)$ of the tangent A_3 is greater than $(1 + a)$. Point P_3 is the position where the tangent has a slope of $(1 + a)$. It is therefore the point corresponding to the second policy (shadow price of S in relation to X is constant): we see that in the case of a resource which renews itself rapidly, this second policy permits withdrawals which are still considerable without putting in danger the possibilities for renewal.

Curve G_2 corresponds to a resource which renews itself less rapidly. Point C_2 corresponding to the first policy permits withdrawals that are still considerable although fewer than for the previous resource. On the other hand, point B_2 is close to A_2: the second policy therefore has two disadvantages: on the one hand the withdrawal that can be used in P_2 is much smaller than in C_2. On the other hand, the danger of

Figure 2

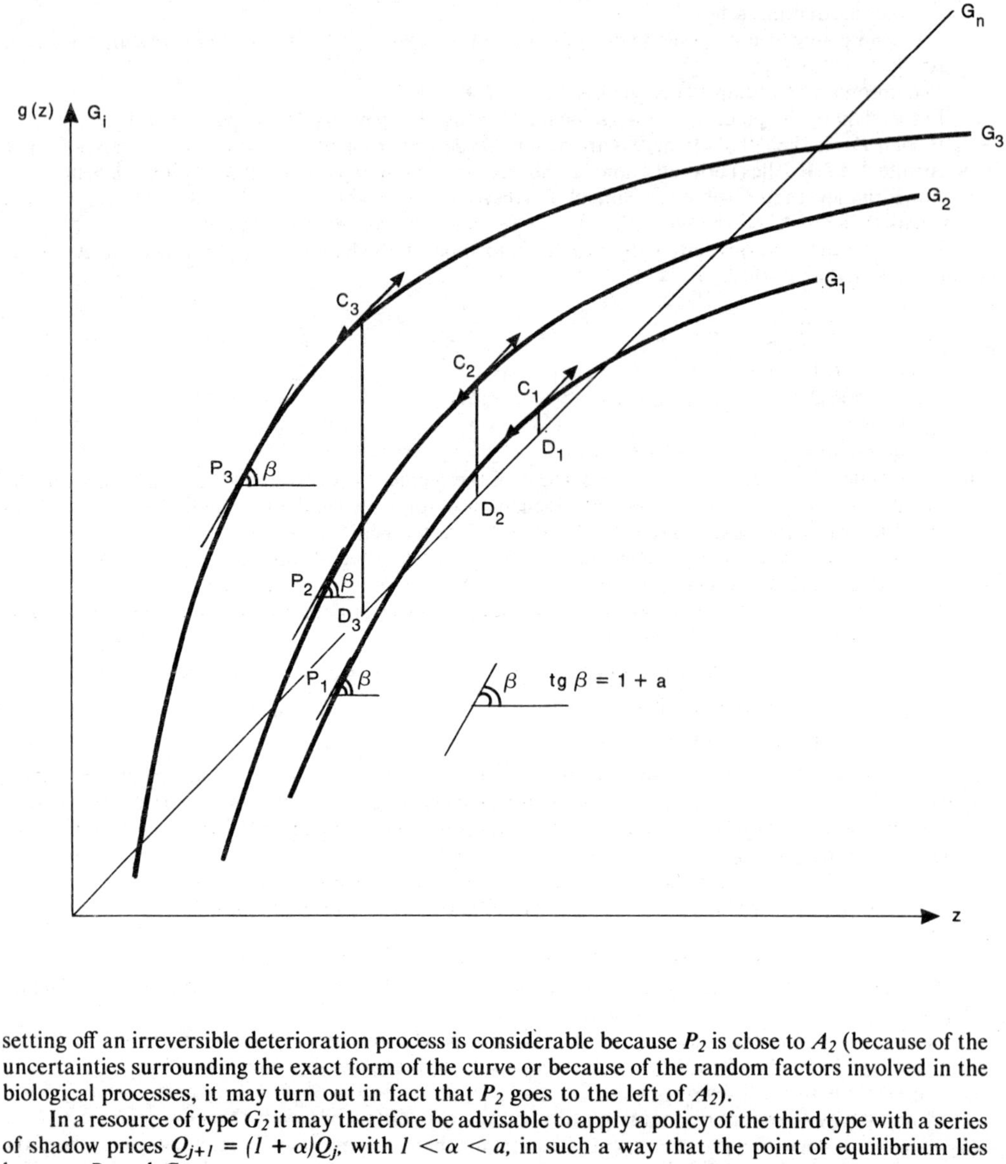

setting off an irreversible deterioration process is considerable because P_2 is close to A_2 (because of the uncertainties surrounding the exact form of the curve or because of the random factors involved in the biological processes, it may turn out in fact that P_2 goes to the left of A_2).

In a resource of type G_2 it may therefore be advisable to apply a policy of the third type with a series of shadow prices $Q_{j+1} = (1 + \alpha)Q_j$, with $1 < \alpha < a$, in such a way that the point of equilibrium lies between P_2 and C_2.

Curve G_1 corresponds to a resource which renews itself very slowly. the first policy (point C_1) permits only a moderate withdrawal but without any great risk. On the other hand, the second policy would correspond to a point P_1 located to the left of A_1 (because we have assumed that the tangent A_1 had a slope between 1 and $1 + a$) and therefore to an irreversible deterioration of the resource.

One is therefore obliged to apply a policy of the third type with a price series $Q_{j+1} = (1 + \alpha)Q_j$ such that:

$$g'(A_1) < \alpha \leqslant a$$

in order that the point of equilibrium lies between A_l and C_l. It is recommended that the choice of α be as near as possible to a, i.e., in fact, that a policy of the first type be applied.

Following discussion of this model we can now come to the following conclusions:

- to exploit a non-renewable natural resource, the only optimal pricing policy is that of assigning a constant discounted shadow price to this resource. Another way of expressing the same policy is to say that it is necessary, in the case of this resource, to assign a relative shadow price which increases at the pace of the discounting rate. (By relative shadow price we mean the ratio of the price of this resource to the average price of goods produced by the economy, according to technological processes which do not involve natural resources);
- to exploit a renewable natural resource it is possible to apply the above policy, and there are two advantages: in a stationary system it permits the exploitation of the maximum quantity of the resource. It carries little risk of irreversibility;
- the policy which consists in fixing constant relative prices (in other words discounted shadow prices which decrease at the same pace as the discounting rate) is possible only when the resource renews itself rather rapidly. In any case, it corresponds to withdrawals on a smaller scale than under the previous policy and to higher risks;
- intermediate pricing policies are possible, with relative shadow prices which increase at varying speeds but at a pace below the discounting rate. The closer the growth rate of the relative prices of the resource is to the discounting rate, the greater will be the quantity exploited and the smaller the risks;
- the longer the period of resource renewal, the more advisable it will be to choose a rate of increase in relative shadow prices close to the discounting rate;
- with regard to forest investments, the longer the period of regeneration of forest populations, the more it will be desirable, in order economically to justify their management, to choose a rate of growth of shadow prices close to the discounting rate;
- one final comment: the whole of the above reasoning implicitly assumes, on the one hand, that man will always need the resource in question and refuse to allow it to be exhausted definitively; and, on the other, that he does not have the technological means of introducing a major change in the shape of curve G, i.e. of making a major change in the rate of the biological processes.

NOTES

1. Special case: If we assume a constant savings ratio denoted by $\gamma = \frac{y_j}{w_j}$ and state that $\frac{x_j + 1}{x_j} = 1 + \eta$ (η is then the growth rate of the economy), it is easy to establish the relationship $(1 + \eta) = (1 + a)\delta$, a formula which links up the growth rate, the discounting rate and the savings ratio.

GLOSSARY

Appraisal

The process of identifying and describing the various effects of a proposed project.

Assessment

Same as "appraisal", but used more narrowly to describe e.g. environmental assessment, social assessment, or financial assessment.

Constraint

An exogenous condition that must be met.

Criterion

A standard against which an effect can be measured.

Distributional effects

Differentiated effects of an action or set of actions on different classes of individuals.

Economic effects

Monetary effects measured in terms of willingness to pay for goods and services, regardless of whether or not such goods and/or services are traded in markets for money.

Economic efficiency

A measure of economic effects indicating the extent to which an alternative represents the "best" solution, economically speaking.

Evaluation

The process of determining to what extent a project alternative satisfies one's objectives, as opposed to simply describing the project effects in an undetached manner.

Feasibility study

An analysis and appraisal of the extent to which a project alternative meets relevant constraints and criteria.

Financial effects

Monetary effects measured in terms of market prices. Effects which cannot be directly or indirectly measured in terms of market prices are not considered in a financial effects analysis.

Goal

A specific condition which is desired to meet an objective, e.g. "our goal is to increase employment and thereby add to the general quality of living".

Indicator

Any dimension along which a social, environmental, economic or financial effect can be assessed.

Market price

The amount of money actually paid to producers for their goods or services.

Multiple purpose project

A project which has more than one goal or more than one use associated with it.

Objective

A general statement of a desirable condition: e.g. "our objective is to improve the general quality of living in area A".

Process

A series of stages involving different steps.

Programme

An interrelated set of projects related to a common objective.

Project

Any set of interrelated actions which are relevant to, and typically take the time of, decision-makers.

Project alternative (or alternative)

One of several means for achieving an objective.

Ranking

The process of ordering alternatives in terms of some effect or order of preference of a person or social group.

Risk

Possible variation in effects which one can describe in terms of probabilities.

Scenario

Description of a set of interrelated actions and their expected effects over time.

Sensitivity analysis

Determining the change in the appraisal of some effect(s) due to changes in a hypothesis.

Shadow price

A monetary value, assigned to a good or service, which expresses the assumed willingness to pay for this good or service (as distinct from what is actually paid for it).

Target

A quantitative measure of a goal, e.g. "our target is to reduce unemployment to five per cent or less in area A".

Uncertainty

Possible variations in outcomes which one cannot describe in terms of probabilities because one does not know the probability distribution.

Valuation

Attaching a monetary value to an effect of an alternative.

BIBLIOGRAPHY

Gregersen, Hans and Contreras, A., 1979, "Economic Analysis of Forestry Projects". FAO *Forestry Paper No. 17*, Rome, FAO.

Gregersen, Hans, 1975, "Effects of Inflation on Evaluation of Forestry Investments". *Journal of Forestry*, September 1975, Washington, D.C., Society of American Foresters.

Gregersen, Hans and Brooks, K, 1980, "Economic Analysis of Watershed projects", FAO *Forestry Paper No. 17, Supplement.*

Hansen, John, 1978, "Guide to Practical Project Appraisal". *Project Formulation and Evaluation Series, No. 3*, UNIDO, New York, United Nations.

Houghtaling, Tomas and Gregersen, Hans, 1979, "Discounting procedures for project appraisals." FAO *Forestry Paper No. 17, Supplement 2*, Rome, FAO.

Inter-American Development Bank, Project Methodology Unit, Country Studies Division. June 14, 1979, "A Methodology for measuring and recording the distributional effects of projects in the Bank's portfolio". Washington, D.C.: Inter-American Development Bank.

Inter American Development Bank, 1977, *Social and Economic Dimensions of Project Evaluation.* Washington, D.C., Inter-American Development Bank.

Little, I.M.D. and Mirrless, J.A., 1974, *Project Planning and Appraisal for Developing Countries*, New York, Basic Books Inc.

Lundgren, Allen, 1973, "Cost-Price: A useful way to evaluate timber growing alternatives". U.S. Department of Agriculture, Forest Service, *Research Paper NC-95*, St. Paul, MN., North Central Forest Experiment Station.

McGaughey, Stephen, 1979, "Investment criteria for agricultural and rural development projects: The measurement of their multiple economic and social consequences", in FAO *Forestry Paper No. 17, Supplement 2*, Rome, FAO.

de Montgolfier, Jean and Bertier, Patrice, 1978, Approche multicritère des problèmes de décision. Editions Hommes et Techniques. Suresnes, France.

Schuster, Ervin, 1979, "Economic impact analysis of forestry projects: A guide to evaluation of distributional consequences." in FAO *Forestry Paper No. 17, Supplement 2*, Rome, FAO.

Squire, Lyn and Van der Tak, H, 1975, *Economic Analysis of Projects*, Baltimore, Johns Hopkins University Press for the World Bank.

UNIDO, 1972, "Guidelines for Project Evaluation". *Project Formulation and Evaluation Series, No. 2*, UNIDO, New York, United Nations.

OECD SALES AGENTS
DÉPOSITAIRES DES PUBLICATIONS DE L'OCDE

ARGENTINA - ARGENTINE
Carlos Hirsch S.R.L.,
Florida 165, 4° Piso,
(Galeria Guemes) 1333 Buenos Aires
Tel. 33.1787.2391 y 30.7122

AUSTRALIA-AUSTRALIE
D.A. Book (Aust.) Pty. Ltd.
11-13 Station Street (P.O. Box 163)
Mitcham, Vic. 3132 Tel. (03) 873 4411

AUSTRIA - AUTRICHE
OECD Publications and Information Centre,
4 Simrockstrasse,
5300 Bonn (Germany) Tel. (0228) 21.60.45
Local Agent:
Gerold & Co., Graben 31, Wien 1 Tel. 52.22.35

BELGIUM - BELGIQUE
Jean de Lannoy, Service Publications OCDE,
avenue du Roi 202
B-1060 Bruxelles Tel. 02/538.51.69

CANADA
Renouf Publishing Company Limited/
Éditions Renouf Limitée Head Office/
Siège social – Store/Magasin :
61, rue Sparks Street,
Ottawa, Ontario KIP 5A6
Tel. (613)238-8985. 1-800-267-4164
Store/Magasin : 211, rue Yonge Street,
Toronto, Ontario M5B 1M4.
Tel. (416)363-3171
Regional Sales Office/
Bureau des Ventes régional :
7575 Trans-Canada Hwy., Suite 305,
Saint-Laurent, Quebec H4T 1V6
Tel. (514)335-9274

DENMARK - DANEMARK
Munksgaard Export and Subscription Service
35, Nørre Søgade, DK-1370 København K
Tel. +45.1.12.85.70

FINLAND - FINLANDE
Akateeminen Kirjakauppa,
Keskuskatu 1, 00100 Helsinki 10 Tel. 0.12141

FRANCE
OCDE/OECD
Mail Orders/Commandes par correspondance :
2, rue André-Pascal,
75775 Paris Cedex 16
Tel. (1) 45.24.82.00
Bookshop/Librairie : 33, rue Octave-Feuillet
75016 Paris
Tel. (1) 45.24.81.67 or/ou (1) 45.24.81.81
Principal correspondant :
Librairie de l'Université,
12a, rue Nazareth,
13602 Aix-en-Provence Tel. 42.26.18.08

GERMANY - ALLEMAGNE
OECD Publications and Information Centre,
4 Simrockstrasse,
5300 Bonn Tel. (0228) 21.60.45

GREECE - GRÈCE
Librairie Kauffmann,
28, rue du Stade, 105 64 Athens Tel. 322.21.60

HONG KONG
Government Information Services,
Publications (Sales) Office,
Beaconsfield House, 4/F.,
Queen's Road Central

ICELAND - ISLANDE
Snæbjörn Jónsson & Co., h.f.,
Hafnarstræti 4 & 9,
P.O.B. 1131 – Reykjavik
Tel. 13133/14281/11936

INDIA - INDE
Oxford Book and Stationery Co.,
Scindia House, New Delhi I Tel. 45896
17 Park St., Calcutta 700016 Tel. 240832

INDONESIA - INDONESIE
Pdin Lipi, P.O. Box 3065/JKT.Jakarta
Tel. 583467

IRELAND - IRLANDE
TDC Publishers – Library Suppliers
12 North Frederick Street, Dublin 1
Tel. 744835-749677

ITALY - ITALIE
Libreria Commissionaria Sansoni,
Via Lamarmora 45, 50121 Firenze
Tel. 579751/584468
Via Bartolini 29, 20155 Milano Tel. 365083
Sub-depositari :
Ugo Tassi, Via A. Farnese 28,
00192 Roma Tel. 310590
Editrice e Libreria Herder,
Piazza Montecitorio 120, 00186 Roma
Tel. 6794628
Agenzia Libraria Pegaso,
Via de Romita 5, 70121 Bari
Tel. 540.105/540.195
Agenzia Libraria Pegaso, Via S.Anna dei
Lombardi 16, 80134 Napoli. Tel. 314180
Libreria Hœpli,
Via Hœpli 5, 20121 Milano Tel. 865446
Libreria Scientifica
Dott. Lucio de Biasio "Aeiou"
Via Meravigli 16, 20123 Milano Tel. 807679
Libreria Zanichelli, Piazza Galvani 1/A,
40124 Bologna Tel. 237389
Libreria Lattes,
Via Garibaldi 3, 10122 Torino Tel. 519274
La diffusione delle edizioni OCSE è inoltre
assicurata dalle migliori librerie nelle città più
importanti.

JAPAN - JAPON
OECD Publications and Information Centre,
Landic Akasaka Bldg., 2-3-4 Akasaka,
Minato-ku, Tokyo 107 Tel. 586.2016

KOREA - CORÉE
Pan Korea Book Corporation
P.O.Box No. 101 Kwangwhamun, Seoul
Tel. 72.7369

LEBANON - LIBAN
Documenta Scientifica/Redico,
Edison Building, Bliss St.,
P.O.B. 5641, Beirut Tel. 354429-344425

MALAYSIA - MALAISIE
University of Malaya Co-operative Bookshop
Ltd.,
P.O.Box 1127, Jalan Pantai Baru,
Kuala Lumpur Tel. 577701/577072

NETHERLANDS - PAYS-BAS
Staatsuitgeverij Verzendboekhandel
Chr. Plantijnstraat, 1 Postbus 20014
2500 EA S-Gravenhage Tel. 070-789911
Voor bestellingen: Tel. 070-789208

NEW ZEALAND - NOUVELLE-ZÉLANDE
Government Printing Office Bookshops:
Auckland: Retail Bookshop, 25 Rutland Street,
Mail Orders, 85 Beach Road
Private Bag C.P.O.
Hamilton: Retail: Ward Street,
Mail Orders, P.O. Box 857
Wellington: Retail, Mulgrave Street, (Head
Office)
Cubacade World Trade Centre,
Mail Orders, Private Bag
Christchurch: Retail, 159 Hereford Street,
Mail Orders, Private Bag
Dunedin: Retail, Princes Street,
Mail Orders, P.O. Box 1104

NORWAY - NORVÈGE
Tanum-Karl Johan a.s
P.O. Box 1177 Sentrum, 0107 Oslo 1
Tel. (02) 801260

PAKISTAN
Mirza Book Agency
65 Shahrah Quaid-E-Azam, Lahore 3 Tel. 66839

PORTUGAL
Livraria Portugal,
Rua do Carmo 70-74, 1117 Lisboa Codex.
Tel. 360582/3

SINGAPORE - SINGAPOUR
Information Publications Pte Ltd
Pei-Fu Industrial Building,
24 New Industrial Road No. 02-06
Singapore 1953 Tel. 2831786, 2831798

SPAIN - ESPAGNE
Mundi-Prensa Libros, S.A.,
Castelló 37, Apartado 1223, Madrid-28001
Tel. 431.33.99
Libreria Bosch, Ronda Universidad 11,
Barcelona 7 Tel. 317.53.08/317.53.58

SWEDEN - SUÈDE
AB CE Fritzes Kungl. Hovbokhandel,
Box 16356, S 103 27 STH,
Regeringsgatan 12,
DS Stockholm Tel. (08) 23.89.00
Subscription Agency/Abonnements:
Wennergren-Williams AB,
Box 30004, S104 25 Stockholm. Tel. 08/54.12.00

SWITZERLAND - SUISSE
OECD Publications and Information Centre,
4 Simrockstrasse,
5300 Bonn (Germany) Tel. (0228) 21.60.45
Local Agent:
Librairie Payot,
6 rue Grenus, 1211 Genève 11
Tel. (022) 31.89.50

TAIWAN - FORMOSE
Good Faith Worldwide Int'l Co., Ltd.
9th floor, No. 118, Sec.2
Chung Hsiao E. Road
Taipei Tel. 391.7396/391.7397

THAILAND - THAILANDE
Suksit Siam Co., Ltd.,
1715 Rama IV Rd.,
Samyam Bangkok 5 Tel. 2511630

TURKEY - TURQUIE
Kültur Yayinlari Is-Türk Ltd. Sti.
Atatürk Bulvari No: 191/Kat. 21
Kavaklidere/Ankara Tel. 17.02.66
Dolmabahce Cad. No: 29
Besiktas/Istanbul Tel. 60.71.88

UNITED KINGDOM - ROYAUME UNI
H.M. Stationery Office,
Postal orders only:
P.O.B. 276, London SW8 5DT
Telephone orders: (01) 622.3316, or
Personal callers:
49 High Holborn, London WC1V 6HB
Branches at: Belfast, Birmingham,
Bristol, Edinburgh, Manchester

UNITED STATES - ÉTATS-UNIS
OECD Publications and Information Centre,
Suite 1207, 1750 Pennsylvania Ave., N.W.,
Washington, D.C. 20006 - 4582
Tel. (202) 724.1857

VENEZUELA
Libreria del Este,
Avda F. Miranda 52, Aptdo. 60337,
Edificio Galipan, Caracas 106
Tel. 32.23.01/33.26.04/31.58.38

YUGOSLAVIA - YOUGOSLAVIE
Jugoslovenska Knjiga, Knez Mihajlova 2,
P.O.B. 36, Beograd Tel. 621.992

Orders and inquiries from countries where Sales Agents have not yet been appointed should be sent to:
OECD, Publications Service, Sales and Distribution Division, 2, rue André-Pascal, 75775 PARIS CEDEX 16.

Les commandes provenant de pays où l'OCDE n'a pas encore désigné de dépositaire peuvent être adressées à :
OCDE, Service des Publications. Division des Ventes et Distribution. 2. rue André-Pascal. 75775 PARIS CEDEX 16.

69777-06-1986

OECD PUBLICATIONS, 2, rue André-Pascal, 75775 PARIS CEDEX 16 - No. 43601 1986
PRINTED IN FRANCE
(42 86 01 1) ISBN 92-64-12826-3